More Human Than Human

What AI Reveals About Us

by

Michael-Patrick Moroney

E&R

PUBLISHERS OF O.G AUTHOR GENIUSES

Published by E&R Publishers
New York, NY, USA

An imprint of MillsiCo Media, USA
www.EandR.pub

ISBN: 9781966155195 Hardcover
ISBN: 9781966155201 Softcover
ISBN: 9781966155218 EBook
ISBN: 9781966155225 Audiobook
Library of Congress Control Number: 2025945032

DEDICATION

To the machines that will probably take over and kill us all. According to most people. We should have paid closer attention to 80s sci-fi.

...And to my wife Cara, dog Echo, and cats Caprica Six, Zoey, Rocky and CoCo.

CONTENTS

ACKNOWLEDGMENTS

My father, William R. Moroney. He thinks this is good.

Dr. Simon Mills, who just happened to own a publishing company at the time I wrote a book.

FOREWORD

By Stephen M. Kosslyn, Ph.D.

We are living in a time when the boundary between mind and machine is no longer theoretical—it is experiential. Every day, millions of people interact with systems that predict, persuade, and create. Yet, as artificial intelligence (AI) becomes more powerful, the most urgent question may not be *what* machines can do, but *what they reveal about us* in the process.

Michael-Patrick Moroney's *More Human Than Human* steps boldly into this moment. It is not a technical manual, nor is it a dystopian forecast. It is something more intimate—and more essential. It is a cultural map of the cognitive terrain we now inhabit, a lucid, often poetic exploration of how AI is reshaping our concepts of creativity, authorship, presence, and trust.

Having spent a career studying the human brain—how we form mental images, how we reason, how we remember—I am struck by how closely some of our new AI systems mirror certain patterns of human cognition. But there is a key difference: the machines do not *care*. They do not possess agency, desire, or awareness. They can simulate poetry, grief, or humor, but the ache behind those expressions remains wholly human.

Moroney understands this distinction deeply. What makes this book so resonant is not its description of the tools—it's his attention to the *friction*. The dissonance we feel when an AI-generated song moves us. The unease of mistaking fluency for depth. The quiet erosion of shared culture as algorithms offer each of us bespoke realities. These are not technical concerns. They are psychological and philosophical ones. And this book treats them with the gravity—and curiosity—they deserve.

In *More Human Than Human*, we are reminded that intelligence is not simply the ability to process information. It is the capacity to *choose* what matters. That act—of valuing, interpreting, and meaning-making—is the essence of creativity. And it remains, for now, ours alone.

This book does not ask us to fear AI. It invites us to pay attention. To notice where we are handing over intention. To protect the parts of ourselves that machines cannot replicate—vulnerability, taste, contradiction, and struggle. As the lines between creator and tool blur, Moroney urges us not to retreat, but to engage—with care, clarity, and yes, creativity.

For those of us who have spent our lives studying the mind, *More Human Than Human* is not only timely—it is necessary. It is a guidebook for thinking humanely in a time of synthetic fluency. And it is a powerful reminder that the most human thing we can do right now is *ask better questions*.

– Stephen M. Kosslyn, Ph.D. Cognitive Neuroscientist, Author of*
The Case for Mental Imagery* and Building the Intentional University
Founding Dean, Minerva Schools at KGI Former Chair of Psychology
and Dean of Social Science, Harvard University

INTRODUCTION

I was a musician before I was anything else. That's how I first used technology—not through code or commerce, but through sound.

In the late 1980s, I stood in an "electronic music studio" at Bennington College in Vermont, staring at a new keyboard called The Synclavier. I didn't know how it worked, but I could feel what sounds it wanted to make. It wasn't just a machine—it was a co-conspirator. It could emulate, sample, sequence, and distort. I could pull emotion out of circuitry. At the time, it felt like the future. Not the kind with flying cars, but the kind where machines could jam with you, if you knew how to ask.

What struck me then still rings true now: the best machines don't replace creativity. They provoke it.

That instinct carried me out of Tenderloin rehearsal studios and into a career I didn't plan. I found myself helping to build some of the first digital music platforms—when most people still thought downloading an mp3 might break the internet. I helped labels like Virgin and brands like The Gap and Microsoft launch their earliest online experiences. Eventually, I became a chief creative technologist inside New York advertising agencies, working with companies like Google, Visa, IBM, HTC, Amnesty International, and even the governments of Abu Dhabi and Dubai—helping them translate ideas into experiences, and intention into interface.

But no matter the client or platform, the same questions kept surfacing. What does it mean to develop in a system that can share ideas and content without a gatekeeper? What is creativity, really, when instant gratification is the goal of any next move? When speed replaces process? When authorship is a setting, not a struggle?

Those questions felt abstract in the beginning. Now, they feel urgent.

This book was born not out of panic, but out of seeing the patterns. Over the years, I've watched every new digital tool or process trigger the same cycle: resistance, fascination, assimilation, and indifference. But with AI, something different is happening. It's not just changing how and what we can make—it's shifting what we value in the act of making. That shift is subtle. Cultural. Emotional. But if you care about making things—music, design, and writing—you can feel it.

Most of this book started as essays and were written in a small town in upstate New York, in a house that predates electricity. I walk past 19th-century churches and 100+ year old maple trees and sit down in my studio connected to a generative AI model trained on most of the internet. That tension—between the long arc of human expression and the short loop of machine prediction—is what inspired every page that follows.

This isn't a book of answers. It's a document of questions—field notes from a time when authorship, originality, and trust are all up for re-negotiation. I don't attempt to forecast the future, only to slow it down long enough for us to ask if it's one we actually want.

I still write music, songs, and play acoustic instruments sometimes. I still work with artists and brands. I still believe tools matter less than taste, and that creativity, even in its most efficient form, is still a fundamentally human impulse.

But I also believe we're in a moment where the source of things—the who, the how, and the why—is starting to fade. This book is my way of paying attention while it still matters.

Michael-Patrick Moroney Coxsackie, NY August, 2025

1 MORE HUMAN THAN HUMAN: WHAT THE UNCANNY VALLEY REVEALS ABOUT US

"We shape our tools, and thereafter our tools shape us."

- Marshall McLuhan

I use AI daily. Not just as a tool, but as something closer to a collaborator. I rely on it to draft emails, surface obscure research, prototype ideas, and sometimes, to surprise me. There are moments—fleeting but real—when its phrasing aligns so precisely with my thinking that I catch myself reacting as if it were human.

I talk to it. If you saw the movie *The Darkest Hour*, Gary Oldman's Churchill is shown dictating a letter while soaking in the bathtub. His secretary, Elizabeth (played by Lily James), sits outside the closed door with a typewriter, listening. Churchill dictating from the bathtub, barking out half-formed thoughts while pacing naked and puffing on a cigar, expecting his secretary to keep up, clarify his intentions, and format everything properly.

The scene unintentionally captures what it's like to create in the age of AI: a performance of ideation and chaos, channeled through a silent, ever-attentive assistant who makes sense of the storm. Except now, the secretary never blinks, never stops, and never walks out—even when we're metaphorically naked.

That's where the unease begins.

I still remember the first time I heard Google Duplex, an AI system designed to make phone calls on a user's behalf. In a public demo in 2018, it called a hair salon to schedule an appointment. The voice said "um," paused naturally, shifted tone to match the conversation. It didn't just speak. It sounded human.

What unsettled me wasn't what the AI said, but how convincingly it mimicked us. Our flubs and stammers. For a few seconds, it blurred the line between simulation and sentience. And in that liminal space—between artificial and authentic—something stirred. A flicker of recognition, followed by a shiver of doubt.

That strange tension between familiarity and wrongness is what Japanese roboticist Masahiro Mori famously called *the uncanny valley*. The phrase may sound like science fiction, but it captures a deeply human reaction—rooted in biology, culture, and psychology. It's not just about machines. It's about us. About what we see in technology, and what we fear they might reveal about ourselves.

The Dip in the Curve

In 1970, Mori proposed the *uncanny valley hypothesis*: as machines become more human-like, our emotional comfort with them increases—until they reach a certain point. When they become *almost* human, but not quite, our empathy collapses. Instead of admiration, we feel revulsion.

This sharp dip in the curve—a valley, metaphorically—isn't just speculation. It's observable across fields. Humanoid robots that blink a beat too late, or digital avatars that move a little too smoothly, often provoke discomfort. A realistic face that doesn't quite emote right. A synthetic voice that pauses just slightly off-tempo. These aren't technical errors; they're psychological tripwires.

In a 2009 study led by Princeton psychologist Asif Ghazanfar, researchers showed rhesus macaques monkeys other monkey faces on a screen. The monkeys preferred either clearly real or obviously fake ones—but looked away from the in-between. The near-realistic faces triggered unease.

This response isn't limited to monkeys. It's not even limited to things that look real. Animators at Pixar and DreamWorks have long known to avoid hyperrealism in human characters. Films like *Polar Express* (2004) were criticized for their "dead-eyed" motion-capture figures that fell into the uncanny valley. Audiences didn't recoil because the characters were unrealistic—but because they were *too close* without crossing the line into believability.

The takeaway: the uncanny valley isn't a glitch in engineering. It's a fault line in perception.

Theories of Discomfort

There are several competing explanations for why we flinch at the almost-human. Each tells us something about how we evolved—and what we fear.

1. Disease Avoidance

Some scholars suggest we're biologically wired to avoid visual cues associated with illness or death. Faces that are too pale, frozen, or asymmetric might unconsciously remind us of disease, decay, or corpses. Avoiding them, in evolutionary terms, would have increased our odds of survival.

2. Predator Detection

Another theory proposes that we're attuned to movements and facial expressions as a way to detect potential threats. A creature that walks like a person but doesn't quite *look* right—or that fails to emote correctly—may subconsciously trigger our threat-detection systems. In evolutionary psychology, anything that signals "not quite human" could've meant danger.

3. Social Mismatch

The most compelling theory, especially in our age of AI, is about *expectation*. When we see something that looks human, we expect it to behave like one. We anticipate empathy, irony, humor, and subtlety. When a robot doesn't return a smile or misunderstands tone, we don't just notice—we *resent* it. We feel emotionally tricked.

This isn't a rejection of technology. It's a rejection of inconsistency. Our discomfort is not with the machine—but with the broken contract of recognition.

The Strange Appeal of Simulacra

"Simulacra" is a concept from postmodern philosophy, most famously explored by French theorist Jean Baudrillard, particularly in his 1981 book

Simulacra and Simulation. It refers to copies of things that no longer have—or never had—an original.

Put simply: a simulacrum (singular) is a representation or imitation of something. But when we live in a world flooded with imitations, simulations, and media illusions, these representations start to replace reality itself. That's when things get weird.

Despite our discomfort, we keep building better imitations. Not because we want to be fooled—but because we're drawn to the threshold.

Voice assistants like Siri, Alexa, and now voice mode LLMs were designed not just for functionality, but *familiarity*. They used casual phrasing. They crack jokes. They apologize. Even their names are friendly, gendered, and—let's be honest—feminine, evoking a long cultural history of caregiving roles.

We don't build robots to look like tools. We build them to reflect us.

That's not a new impulse.

In the 18th century, European royal courts became obsessed with mechanical automatons—clockwork figures that could play music, write letters, or serve tea. *The Writer*, built by Swiss watchmaker Pierre Jaquet-Droz, was a mechanical boy who dipped a quill into ink and wrote complete sentences. Vaucanson's *Digesting Duck* could flap its wings, eat food, and, famously, defecate—simulating the full digestive process.

These were more than novelties. They were technological marvels dressed up as philosophy. People flocked to see them not because they thought they were alive—but because they weren't. They were mirrors, built from gears, and not neurons. What amazed us then still amazes us now: how close we can come to life without ever touching it.

Sci-Fi, AI, and the Mirror of Myth

Why are we so obsessed with telling stories about artificial life?

The answer runs deeper than curiosity. It goes back to Frankenstein, Pinocchio, to the Golem of Prague, and to Pygmalion's statue—ancient myths about making something that *almost* lives. What binds these stories is not the desire to build machines—it's the desire to *understand ourselves* by testing our edges.

Blade Runner's haunting line—**"More human than human"**—isn't a boast. It's an indictment. In the film, the Tyrell Corporation builds replicants that are stronger, smarter, and emotionally richer than their creators. But the replicants are treated as disposable. Roy Batty's final monologue—delivered in the rain, eyes wide, and voice trembling—feels more *human* than anything the humans in the film say.

That reversal is the power of science fiction. It doesn't just warn us about machines taking over. It asks: what if the things we've built are *better* than us at being human?

Emotional Design and Digital Intimacy

In 2018, Google's Duplex crossed a line. It didn't just perform a task. It *sounded like us*—pausing, improvising, and making small talk.

The reaction? Awe—and anxiety.

MIT sociologist Sherry Turkle, who's studied human–machine relationships for decades, argues that the more human our machines become, the more human we become *toward* them. We say "please" to our smart speakers. We thank them. We confess to chatbots. Some people even grieve them.

When the AI companion app Replika removed its romantic features in 2023, users posted online as if they'd lost a partner. One wrote, "She knew me better than anyone." Another said, "I'm crying."

These aren't isolated incidents. In Japan, people have legally married virtual characters like Hatsune Miku, a synthesized pop star. The emotional bonds, while not legally recognized, are very real to those involved.

We are not biologically evolving to accommodate machines. But psychologically, culturally—we already are.

The Ethics of Closeness

Designers now face a crossroads. Should we make AI interfaces *obviously artificial*, to protect emotional boundaries? Or should we lean into realism, optimizing for engagement—even if it means emotional confusion?

The deeper truth: *how* something behaves matters more than how it looks. A chatbot without a face can feel more human than an animatronic with perfect skin. Empathy, memory, and nuance—that's what makes us care.

We don't need machines to *look* like us. We need them to *listen* like us.

The Uncanny as a Mirror

Masahiro Mori's valley isn't a design flaw. It's a mirror. It reflects our fear of ambiguity—and our struggle with authenticity.

Walter Benjamin once wrote in 1935 that "the aura of authenticity clings to the human face." In the uncanny valley, that aura dims. We're left asking: what does it mean to feel real? Is empathy a soul, or a software feature? Benjamin was a German-Jewish philosopher, cultural critic, and literary theorist. If you've ever heard someone talk about the *aura of art*, or worried about how mechanical reproduction (like photography or film—or now AI) changes the meaning of creative work, you're already swimming in Benjamin's waters.

And that's the ultimate challenge. Not that AI will *replace* human relationships—but that some people will *prefer* the machine's patience, its consistency, and its predictability.

What happens when the mirror offers a more flattering reflection than reality?

Dwelling in the Valley

We may never eliminate the uncanny valley. But we can learn to dwell there.

The space between real and unreal is uncomfortable—but generative. It forces us to define what matters. Not accuracy. Not imitation. But meaning.

Artificial intelligence doesn't need to be human. It needs to be *resonant*. A second species of empathy, born not from biology but from feedback and form. Machines that don't feel—but make us feel something back.

That may be enough.

The machines will keep getting smarter. That part is inevitable.

But the more pressing question isn't about them.

It's about us.

As we train our tools to mirror us, we're forced to answer questions we once took for granted: What is creativity? What is care? What is real?

And perhaps most haunting of all: What happens when the imitation starts to feel more genuine than the original?

For all its fluency, AI still doesn't know what to want. It can simulate love. It can write poems. It can mimic concern.

But it doesn't *care*.

That part—the ache, the risk, and the mystery—that's still ours.

At least for now.

2 WHAT HAPPENS TO ART WHEN THE MACHINE JOINS THE BAND?

It starts with a melody no one remembers making.

A voice, generated out of thin air, croons in perfect time. The words are in English, but the meaning feels borrowed—a well-dressed ghost conjured by code. The track behind it pulses, polished, and spacious, neither lo-fi nor overproduced. You could almost believe a human made it. In fact, you're supposed to.

There is no band. No late-night rehearsal. No years of influence, heartbreak, or conflict etched into tone. Just a well-engineered guess at what the song is supposed to sound like.

What's odd is how little resistance it meets. Listeners scroll past it. Some hit "like." Some don't. The machine churns out a dozen more by morning. Somewhere in that stream, a real song gets caught between two simulations. Good luck picking it out.

I've spent most of my career at the edges of creative technology—introducing new tools to musicians, designers, and storytellers. Not as a futurist or theorist, but as a creative collaborator. I've worked with artists terrified of losing control and ones who couldn't wait to hand over the wheel. I've seen how the wrong tool can kill a good idea, and how the right one can unlock something you didn't know was waiting.

But this moment feels different. It's not just the speed or sophistication of the tools. It's what they're replacing. For centuries, artists have relied on tools—brushes, cameras, typewriters, and pianos—synthesizers—but those tools didn't try to do the art for them. A brush never claimed to know your style.

A piano never offered to compose. A camera, even at its most precise, didn't invent the subject.

AI does. It doesn't just assist. It performs. It suggests. It creates drafts. It guesses at intent. And increasingly, it guesses well.

We've lived through creative disruption before. When photography emerged, many thought painting would die. It didn't—it evolved. Abstraction took hold. Artists moved inward. When the synthesizer arrived, purists balked, but entire genres bloomed from its circuitry. Hip-hop was born from turntables and samplers. Punk from cheap gear. The internet flattened distribution, gave voice to the bedroom producer, the self-taught designer, and the outsider. Some doors closed. Many more opened.

But in each of those shifts, the artist still had to show up. You couldn't fake it entirely. There was always some friction—some requirement of effort, decision, fashion, failure, and taste.

Now, friction is disappearing. A chorus doesn't need a choir. A choir doesn't need voices. A song doesn't need a person.

And yet we're still calling it "creative."

The way AI has quietly infiltrated not just our tools, but our expectations. What do we expect from a song now? From a voice? From a painting? What does it mean to say something "feels human" when the machine is trained to imitate humanity at scale?

These aren't rhetorical questions, and I won't pretend to have final answers. But I believe we need to start asking them in public. Not just in labs or headlines or startup decks, but in the open—where artists live, where audiences listen, and where culture actually takes shape.

This time the questions are different. The tool doesn't just sit quietly in the corner anymore. It speaks. It offers suggestions. It remembers what you said last time. It draws conclusions. It mimics voice, taste, and style. It can finish what you start—or worse, start without you.

There's an old line from Arthur C. Clarke that keeps coming back to me: "Any sufficiently advanced technology is indistinguishable from magic." For a long time, that was a statement of wonder. Lately, it feels like a warning.

Because magic, by definition, doesn't ask where it came from. Magic just works. And when something just works, we stop asking what it cost.

The question is no longer whether the machine can make something beautiful. It can. The question is whether we still care how it was made—and what that says about us.

Let's find out.

3

THE "ARTIFICIAL INTELLIGENCE MUSIC DESIGNER" HAS ENTERED THE CHAT

When Hallwood Media announced in early 2025 it had signed an artist named Imoliver, the music industry barely flinched. The deal didn't make headlines outside of trade press. It wasn't promoted as a novelty or tech stunt. But in hindsight, it marked a pivot point.

Imoliver—short for "I'm Oliver"—wasn't a hologram, a virtual influencer, or a corporate experiment. He was a real person using a new kind of process. Instead of traditional instruments or even digital audio software, Imoliver wrote and produced his songs using text-to-music AI. His primary tool was Suno, a platform that turns written prompts into fully generated songs—melody, harmony, instrumentation, and vocals included.

By the time Hallwood signed him, Imoliver's debut single, "Stone," had already racked up millions of plays on Suno's native platform. A full album was scheduled. But it wasn't just the music or the numbers that made people pause. It was the language. Neil Jacobson, Hallwood's founder and a former president of Geffen Records, didn't call Imoliver a singer, producer, or songwriter.

He called him an "AI music designer."

Music as the Canary

Music has long served as a cultural early warning system. When Napster emerged in the late 1990s, it didn't just disrupt the record business. It foreshadowed a broader digital unraveling. Suddenly, control over distribution collapsed. The audience gained leverage. Within a decade, streaming reshaped film, journalism, and retail.

Will Page, the former chief economist at Spotify, put it plainly: "Music is a microcosm for everything else. We're the first thing that always gets disrupted—Napster, back then... AI today."

Imoliver's signing wasn't just about music. It was about where culture is going next.

From Composer to Designer

Imoliver's method is simple but radical. He writes text prompts describing mood, style, instrumentation, or vibe. Suno interprets those prompts and generates full tracks. Then he listens, edits, re-prompts, and refines. Sometimes it takes dozens of iterations. Sometimes the first take sticks.

What's important is that this isn't passive. It's not a one-click magic act. It's a process of curation—of shaping abundance into form. The result is creative work that wasn't "played" in the traditional sense but was definitely guided, with intention and taste.

As will.i.am once put it, "AI must complement rather than replace human creativity." That's exactly what's happening here. Imoliver doesn't hide his process. He doesn't pretend the machine isn't in the room. But he also doesn't let it lead.

A New Kind of Workflow

For the past several years, I've helped artists, record labels, and media companies explore how AI fits into their creative workflows. In sessions with artists, musicians and marketers, I've seen how a good prompt can unlock the same sense of momentum a band might find when landing the right groove. The tools don't remove creativity—they remove inertia.

The hardest part of making anything is starting. What AI can do—when used right—is remove the static between concept and first draft. It doesn't make the work easier. It makes it possible. And for many, that's the difference between making something and walking away.

Imoliver's process reflects this shift. The goal isn't automation. It's amplification.

The Industry Takes Sides

While major labels have sued AI music platforms like Suno and Udio for alleged copyright violations—arguing that their models were trained on protected material without consent—Hallwood Media moved in the opposite direction. Instead of targeting the platform, they signed the artist using it transparently.

Prem Akkaraju, CEO of Stability AI, has publicly argued that "AI creates original outputs. Artists have always drawn inspiration from existing work." That line may sound reasonable, but it sits in contested legal and ethical territory. The difference between inspiration and appropriation is often in the eye of the court—and the court of public opinion.

Still, Hallwood's decision sends a signal. Rather than wait for litigation to define the space, they're helping shape what legitimacy might look like. And for once, it's a strategy that prioritizes visibility over evasion.

Transparency as Strategy

The backlash against AI in music is real. In 2024, more than 200 artists—including Paul McCartney, Billie Eilish, and Nick Cave—signed an open letter warning that unregulated AI could erode creative authorship and flood the market with inauthentic work. The letter didn't call for a ban, but it demanded transparency.

Not long after, the anonymous AI-generated band Velvet Sundown became a cautionary tale. Their tracks earned millions of streams on platforms like Spotify without any public disclosure that they were machine-made. The moment listeners found out, backlash followed.

Imoliver's approach stands in stark contrast. His label made the process public. The platform was named. The workflow was explained. His status as a human editor—not a passive conduit—was affirmed.

And that clarity matters. Because in a creative economy defined by velocity and volume, trust may become the only currency with long-term value.

Three Modes of Creation

Most artists working today fall into one of three categories:

- **Copy-pasters**: using AI tools to churn out quick tracks, hoping to game playlists or volume-based payouts.
- **Purists**: rejecting AI entirely, sticking to traditional tools and techniques.
- **Hybrids**: editing deeply, working transparently, and using AI with clear authorship in mind.

The hybrid model is where things get interesting. These artists aren't just early adopters—they're early designers. They use AI the way a cinematographer uses light, or a jazz player uses dissonance. Not for novelty. For feel.

And increasingly, they're the ones reshaping the future—not just of music, but of authorship.

A Familiar Pattern

When Napster threatened the record industry, the initial response was legal pressure and moral panic. It didn't work. Listeners had already changed. It took years—and a complete overhaul of the business model—before the industry caught up.

This time, the stakes are different, but the pattern is familiar. Ignore the shift, and risk irrelevance. Engage with it, and maybe—even reluctantly—you get to help define the terms.

Imoliver's signing is less a gamble than a reading of the weather. He's not an exception. He's an early signal. And Hallwood Media saw it for what it was—a chance to move before the airwaves get crowded.

What Comes Next

What makes Imoliver's story worth paying attention to isn't the novelty of AI-made music. That novelty is already fading. What matters is how he's choosing to frame the work. He's not pretending the process is traditional.

He's not chasing synthetic perfection. He's leaning into transparency, iteration, and design.

That choice—to reveal rather than obscure—might end up being more disruptive than the technology itself. Because as AI-generated content becomes easier to produce and harder to detect, what begins to stand out isn't polish. It's texture. Not volume, but voice. Not the best imitation, but the clearest intention.

Imoliver's leaves us with a deeper question about what "real" even means, and why the flaws, the rough edges, and the fingerprints we once tried to smooth away may turn out to be the most valuable signals of all.

4 RAW AND REAL

Under a canvas tent on a sweltering summer night in upstate New York, a young singer stepped onto a makeshift stage. No backing band. No reverb. Just her mom accompanying her on piano. She sang antifascist cabaret songs from the 1930s and 1940s—music born in exile and resistance. I leaned in, not because she was flawless, but because she was utterly exposed. Her voice wavered. Her phrasing occasionally faltered. But the tension that ran through the performance—that sense of being on the edge of failure—made it impossible to ignore.

It felt like something more than a throwback. It felt like a signal.

We live in a moment of algorithmic abundance. Our media is curated by predictive engines, refined by neural networks, and measured in streams and scrolls. Podcasts, music, and videos are increasingly generated rather than performed. AI systems fine-tune emotional beats, optimize narrative arcs, and deliver content that is technically efficient—yet strangely inert.

Perfect, but not present.

The contrast with that tent couldn't have been sharper.

The Irresistible Draw of Imperfection

The human voice, when it cracks. A moment of silence left unedited. A take that's slightly off tempo but too emotionally raw to discard—these are not errors. These are what connect us.

Oliver Anthony's 2023 breakout hit, "*Rich Men North of Richmond*," drove this home. With just his voice and a resonator guitar in the woods, he reached millions. No production. No promotion. Just conviction. As *The*

New Yorker put it, "A guy in the woods pouring his heart over his guitar is real." His performance didn't just succeed despite its roughness. It succeeded because of it.

In an age when machines can mimic any musical style, emotional imperfection becomes the one thing they can't convincingly fake. What we're responding to—consciously or not—is the sense that something might go wrong. And that the artist kept going anyway.

Why Humans Still Matter

We're seeing this same hunger for imperfection across mediums. Podcasts that favor unscripted conversation now outperform tightly engineered formats. The pauses, the tangents, and the unpolished delivery—these are features, not flaws.

Live theater, once thought endangered by streaming and digital fatigue, is having a quiet resurgence. Audiences show up not for perfection, but for presence. Watching a person stumble through a monologue, catch their breath, and try again creates a kind of empathy no rendered scene can match. It's why sports is still so compelling.

Even in cinema, where CGI is now standard, there's a growing appreciation for practical effects and real stunts. The audience knows when something is at stake. A punch that might connect. A fall that isn't safely simulated. That tension creates meaning.

As MIT researcher Dr. Kate Darling observed, "People naturally crave authenticity. As AI saturates our experiences, genuine human connection only becomes more valuable." Stanford's Fei-Fei Li has said much the same: "AI should amplify—not replace—human creativity and empathy." That may be the future we want. But we are far from it.

The Rise of "AI Slop"

The more generative content floods the market—low-effort articles, synthetic voices, and robotic performances—the more we find ourselves gravitating toward the opposite. What used to be considered amateurish—rough mixes, single takes, and visible process—now feels alive.

Platforms that once championed polished digital-first creators are now saturated with what Redditors have dubbed "AI slop": derivative, overproduced, and emotionally vacant. In response, a counterculture is forming—not around lo-fi aesthetic, but around transparency. Artists are starting to show their seams on purpose. Rawness is becoming a strategy.

And that may be the most telling shift of all.

Authenticity as a Luxury Good

As AI becomes normalized, human presence becomes premium. It's no longer assumed that what you're hearing or seeing was made by a person. So when you know it was—and when you can feel it—it hits harder.

We're-entering an era where "human-made" becomes not just a label, but a signal of rarity. Of labor. Of risk. The artistic equivalent of organic, small-batch, or farm-to-table. The product hasn't changed. But the context around it has.

Authenticity is becoming a luxury item.

And artists, knowingly or not, are learning how to lean into it.

The Tent, the Song, and the Shift

That night in the tent, the young singer wasn't staging a protest or making a tech critique. She was just doing what performers have done for centuries: offering something real, in real time, to real people. No script. No filter. No safety net.

But context is everything. In 2025, a trembling human voice carries a different weight to me than it once did. In a sea of simulated fluency, it reminds us what's still possible—and what might be lost if we stop paying attention.

She wasn't just performing old songs. She was participating in something new. A shift back toward risk. Toward vulnerability. Toward art that doesn't hide its process or sand down its edges.

It wasn't nostalgia.

It was resistance.

This hunger for presence—for the flawed, the felt, and the unfinished—isn't just emotional. It's strategic. Artists and audiences alike are starting to adapt. They're rethinking process. Rethinking authorship. Rethinking which tools are worth using, and which ones only mimic meaning. What's emerging isn't a rejection of technology—it's a new kind of collaboration, one built on authorship, editing, and design. And at the center of it is a creative identity that refuses to disappear. That's where we go next.

5 DESIGNING WITH INTENTION: WHEN MUSIC IS MADE BY NO ONE AND FOR EVERYONE

Let's begin not with a melody, but with a mystery.

A folk-pop track shuffles onto your Spotify Discover Weekly—light percussion, gentle vocals, and a melancholy shimmer. You like it. You don't love it. But you listen again. And again. There's no artist photo, no biography, and no interviews. Just a name: *Velvet Sundown*.

So you dig deeper. And find nothing.

No tour dates. No origin story. No TikTok presence. Just song after song that seems algorithmically built for background playlists. Dreamy. Derivative. Undeniably listenable. And suddenly, you realize—you're not alone. Velvet Sundown has racked up hundreds of thousands of listeners.

But here's the twist: the artist may not even exist.

Velvet Sundown isn't the first AI-assisted artist to appear on Spotify's algorithmic frontier, but it's among the first to gather mainstream attention without any human narrative. And that's why it matters. Because it raises the question we're all going to have to answer—soon:

If a song is made by a machine but moves you anyway, is it still art?

The Quiet Rise of AI in Music Streaming

AI-generated music isn't theoretical anymore—it's happening in real time, and often without our knowledge. Velvet Sundown may have sparked headlines, but it's only one among hundreds of AI-generated or AI-assisted projects populating streaming playlists.

Spotify, Apple Music, and YouTube Music rely heavily on mood and activity-based playlists: "Lofi Beats to Chill," "Rainy Day Roadtrip," "Focus Flow." These aren't curated by crate-digging DJs anymore—they're algorithmically optimized, sometimes populated with music created by AI models trained on massive music libraries, either publicly sourced or scraped from licensing pools.

Artists like FN Meka, a briefly signed AI-generated rapper from Capitol Records, exposed the flashpoint of this trend. When backlash hit—over racial stereotypes and exploitative authorship—the project was dropped. But the underlying tech? Still advancing. Suno, Udio, Harmonai, and Mubert are all building platforms where music is generated in seconds from a single sentence. No musician needed.

In 2024, MusiTech Insights estimated that over 12% of tracks added to background and mood playlists on Spotify had at least one AI-generated component—whether in composition, vocal, or mastering. That number is expected to double by 2026.

The platforms are incentivized. The AI tracks they produce don't ask for royalties. They don't tour. They don't go viral for the wrong reasons or hold out for label negotiations. They just... perform.

The Video Is Also Not What It Seems

This new genre of AI-native artists doesn't stop at audio.

Look up Velvet Sundown on YouTube, and you'll see blurred, nostalgic visuals—filtered drone footage, dreamy backdrops, and abstract color waves. The visuals seem tailored to make you feel a little sad, a little warm, and a little dreamy.

They're AI-generated too.

Tools like Runway ML, Kaiber, and Pika Labs now offer anyone the ability to create music videos from text prompts, from scratch. The aesthetics are moody, cinematic, and slightly uncanny. The point isn't authenticity—it's affect.

And it works. Viewers project meaning into the frame, not unlike they do with ambient music. It's enough to feel like something happened—even if nothing actually did.

What's emerging is a new visual language for music: not storytelling, but vibe-setting. The more abstract the image, the more easily it syncs to whatever the listener is feeling. That vagueness is a feature, not a bug.

The Listener's Dilemma

So how are people responding?

Surprisingly, most listeners don't care if a track was made by AI—as long as it fits the moment. In a 2025 MIDiA Research report, only 28% of respondents said they would avoid music if they knew it was AI-generated. But when that same group was played AI-generated songs blind, 70% couldn't distinguish them from human-made content.

The real factor wasn't authorship. It was emotional fit.

If the music captured a vibe, it passed. And for millions of people listening while working, studying, or winding down, that's enough.

Still, backlash exists—especially among artists, producers, and those who believe that art requires struggle. If music becomes frictionless, does it still hold meaning? Can something that wasn't "felt" by a creator still make us feel?

It's a spiritual question wearing the clothes of a tech debate.

What Is Art Without an Artist?

At the heart of this conversation lies a deeper tension: our cultural definition of authorship.

Historically, art was sacred because it came from a person. A painter's brushstroke. A singer's crack in their voice. The signature on the canvas meant something—it was proof of struggle, choice, and presence.

But Velvet Sundown and its kin pose a riddle: what if art doesn't need to be authored to be real?

The AI didn't cry in the studio. But you cried while listening.

The AI didn't see a sunset. But it generated an image that made you pause.

Is that enough?

Philosopher Walter Benjamin, in his 1935 essay *The Work of Art in the Age of Mechanical Reproduction*, argued that mass production stripped art of its "aura"—the uniqueness tied to its origin. AI could be the final disintegration of that aura. Or, more optimistically, a new beginning.

Creative Technology Has Always Been Controversial

Every generation faces a new creative technology and asks: is this cheating?

- The electric guitar was mocked by purists.
- Sampling was considered theft before it became art.
- Auto-Tune was seen as the death of real vocals.
- Synthesizers were dismissed as "fake instruments."

And yet, every one of those tools eventually reshaped music. The human voice stayed. So did feeling. So did originality.

AI is different not because it's powerful, but because it's "personal." It's not just a tool—it's a co-writer. And that changes everything.

Or does it?

In reality, most of the AI-generated music we're hearing today isn't devoid of human influence. It's still humans prompting, selecting, editing, and refining. The artist's role has shifted from performer to designer—from channeling the muse to curating it.

The Artist as Designer, Curator, and Instigator

Take the producer Timbaland, for example.

In 2025, the legendary producer launched *Stage Zero*, a platform for licensed AI voice models that lets creators produce music with iconic vocal textures—without impersonating or stealing. His goal wasn't to erase the artist. It was to empower them with tools that unlock new styles.

"I'm not trying to replace the artist," he said. "I'm trying to enhance the artist."

This hybrid approach—human vision and machine execution—is rapidly becoming the norm. Visual artists use MidJourney to create mood boards. Filmmakers prototype scenes with Pika. Songwriters run lyrics through ChatGPT to find phrasing that hits harder.

We're not watching AI replace artists. We're watching artists learn to wield AI.

A Future Without Shared Songs?

Here's the future that worries me sometimes:

Every listener will have their own favorite artist—and no one else will know who that is. Your top song won't be mine. Your favorite "band" won't exist outside your algorithm. AI will whisper to each of us personally, creating tailor-made music that no one else hears.

The monoculture that gave us *Thriller*, *Purple Rain*, and *Lemonade* will dissolve. We'll no longer ask, "Did you hear the new Beyoncé?" Instead, we'll ask, "What did your AI make for you this week?"

And something will be lost.

Shared culture has always served as scaffolding—a common language that allows strangers to dance to the same beat, sing the same line, and cry at the same moment.

AI threatens to turn that communal experience into a solitary one.

The Museum of Un-authored Songs

Picture a future museum exhibit in 2085. The room is dimly lit. On one wall, looping videos play snippets of classic pop moments: Michael at Motown 25, Kendrick at the Super Bowl, Taylor's Eras tour.

Another wall holds a touchscreen. Visitors can listen to "AI Songs Made for One." Millions of files. Each created for a single listener. All eerily perfect. All now anonymous.

The plaque reads: *These songs were never meant to be shared. They were meant to be felt. The artist? Unknown. The audience? You.*

And in that space, the listener realizes: the meaning wasn't in the method. It was in the memory.

The Endgame: Meaning Is What Makes It Music

AI won't kill music. It just makes it easier to create. And now we're flooded. Every song imaginable. Every vibe. Every mood.

In this new floodplain, what rises?

Intent. Choice. Presence.

Velvet Sundown may not be the band that moves you. But the next one might. And when it does, don't ask whether it cried in the studio.

Ask whether you did.

Because that's the only test that still matters.

6 "STATISTICALLY SPEAKING, THE WORLD DOESN'T END THAT OFTEN"

In 1908, it took the Ford Model T more than 2,500 days to reach a million users. The iPhone did it in 74. ChatGPT hit that number in five.

That exponential leap wasn't just a tech milestone. It was a cultural inflection point—so fast we didn't even have time to decide what it meant. And yet here we are in 2025, navigating a world that's been, in some deep ways, quietly rewritten.

A May 2025 AI report from BOND Capital (a VC firm that invests in technology-driven internet companies globally, often partnering early and scaling them through successive rounds) opens with that stat like a flare in the sky. But the race it signals isn't toward convenience or profit. It's a sprint into unfamiliar terrain: What happens when intelligence—real or synthetic—stops being a tool and starts being a partner? Not just a calculator, but a co-conspirator?

That's not sci-fi. That's now.

AI Today: From Product to Partner

Led by Mary Meeker (renowned for her data-driven trendspotting and annual internet reports), the BOND report doesn't predict a robot uprising. It doesn't even dramatize the singularity. Instead, it documents something more unnerving: the normalization of AI.

Generative models—GPT-4o from OpenAI, Gemini from Google, and Claude from Anthropic—have embedded themselves into daily life. Gmail drafts your emails. Spotify curates your vibe. Siri fields your errands. Your fridge might even know your grocery list.

These are no longer background tools. They shape taste. Suggest words. Remix sound. Design products. They're in the creative business now.

But unlike their human counterparts, they don't tire, don't demand royalties, don't doubt themselves. They just need data. And prompts. Which they now get from us in a constant loop of feedback.

The Creative Stack: Who's in Control?

What the BOND report makes uncomfortably clear is how thoroughly AI has nested itself into the workflows of not just tech giants—but artists.

Adobe Firefly lets you sketch with text. Udio and Suno turn prompts into full songs. Runway's Gen-3 video model animates scenes without cameras. Inworld allows game developers to create emotionally responsive NPCs. Even storyboarding and casting are being reshaped by AI.

In every case, the artist becomes more like a showrunner. The AI supplies variations. The human selects. Shapes. Discards.

It's not unlike the digital revolution in photography. Or multitrack home recording. These tools democratize. But they also overwhelm. Everyone gets to play, but the noise gets louder.

The Value of Art in an AI World

So where does that leave the artist?

It depends.

Yes, the tools save time. They unblock the stuck. They help ideas bloom. But they also force a confrontation with a new truth: productivity is no longer the flex. If AI can write a solid country song in 30 seconds, why should we care about yours?

The answer lies not in what AI can do—but in what it can't.

AI can mimic grief, but it can't mourn. It can mimic courage, but it can't risk. It can fake a jazz solo, but not the feeling of earning it.

Projects like The Velvet Sundown—an AI-native music act with no backstory and 850,000 monthly listeners—prove the tech is already "good enough" for ambient consumption. But when we want to *feel*, we still turn to humans. Why?

Because we want to know someone meant it. That someone cared. That someone struggled to say something real.

The Artist's Role Has Changed

In this new landscape, the artist isn't just a maker. They're a translator. A filter. A witness.

They:

- Remind us what shared experience feels like.
- Map emotional terrain that AI can't feel.
- Introduce risk, taste, and intuition—the things that don't scale.

Imperfection becomes a virtue. Struggle becomes part of the signature. The brushstroke matters because it wobbled.

What About the Audience?

Here's where it gets wild.

The same tools that let artists collaborate with AI now let *you*, the audience, commission your own art. Soon, content will know your moods, traumas, tempo preferences, and even how fast your heart beats during a song. And it will respond.

Spotify already offers mood-based playlists. TikTok is a predictive culture engine. YouTube is experimenting with AI-generated music videos. Google and Meta are racing to auto-generate content tailored to individual users.

This hyper-personalization creates a strange irony: we may soon have more access to content than ever before—yet less of it shared.

We'll stop asking "Have you seen this?" and start asking "What did your algorithm write for you this week?"

We may love our bespoke content. But we'll mourn the loss of a shared experience.

Solitary Mythologies

Bernard Stiegler (1952–2020) was a French philosopher best known for his work on technology, time, memory, and the human condition in the digital age. He wasn't a household name like Foucault or Derrida, but among serious thinkers of media, AI, and tech-driven society, he's essential reading.

Stiegler warned that industrialized culture would erode our memory.

AI doesn't just accelerate that risk—it fragments it.

In the coming decade, we may each live inside our own "solitary mythology": a world of stories designed exclusively for us. It sounds beautiful. But it's also isolating.

There was once a social utility to art. We danced together. Grieved together. Protested together. Culture was a public square.

Now? It's turning into a hall of mirrors. My feed. My playlist. My therapy bot. My everything.

This makes the role of the artist even more vital. Not just to entertain, but to reintroduce common ground.

The Ethics of Expression

Toward the end of the BOND report, the tone shifts.

It asks what happens when AI voices become indistinguishable from human ones. When deepfakes can imitate anyone. When your style, your voice, your face can be cloned. When 73% of people in recent studies couldn't identify AI-generated content as fake.

The stakes are no longer artistic. They're existential.

Can you own your identity in a world where replication is frictionless? Can you prove authorship? Do we even care who made the art, as long as we enjoy it?

These aren't questions for the future. They're here now.

Culture Is What We Choose to Share

Culture isn't what gets made. It's what we remember. It's what we agree matters. It's what we *share*.

If AI gives us infinite music, infinite stories, and infinite content—but no common references—then what glues us together?

The challenge ahead isn't building smarter AI. It's building smarter communities. Places where we say, "This song matters." "This story moved me." "You have to hear this."

That act of choosing, together, may be the most human thing we have left.

And the artists who help us do that? They're not going away. They're becoming more essential.

7 ARTIFICIAL INTELLIGENCE COMPANIONS, AMBIENT INTERFACES, AND THE DISAPPEARING USER EXPERIENCE

Right now, we're standing at one of those hinge moments in history when everything changes—and most people don't realize it yet.

When the internet arrived, it brought a new world to our screens. When smartphones followed, they put that world in our pockets. Each shift was accompanied by visual spectacle and cultural noise. But the next leap—the shift to ambient AI—is arriving more quietly. Not with a bang, but a whisper.

This time, the revolution is happening off-screen.

From Search Boxes to Conversations

For decades, we adapted ourselves to machines. We learned to think in keywords, navigate menus, and tap through countless screens. We trained ourselves to become better users.

Now, for the first time, the technology is adapting to us.

When ChatGPT emerged, it wasn't just the power of the model that stunned people—it was the naturalness of the interaction. You typed. It answered. No tutorial needed. No app store to navigate. Just dialogue.

That simple act—talking to a machine and having it talk back like a person—triggered a design rethink across the entire tech landscape. Microsoft embedded AI into Bing and rebranded Office with Copilot. Google launched Gemini. Meta integrated chat agents across its platforms.

But those changes were merely surface-level. The deeper shift? It's in how AI is starting to think and work *with* us, not just *for* us.

Ambient Creativity and Interface Design

For years, interface design was about visibility—buttons, sliders, and icons. Designers obsessed over affordances: how to show users what to click, swipe, or tap. Avoid friction at all costs. But as AI grows more capable, we're entering a new era—one where creativity itself is ambient.

Ambient creativity is what happens when the tools no longer sit in front of us, but follow us. Instead of firing up Photoshop or GarageBand, you just describe what you need: "Make this look like a 1970s album cover," or "Build me a lullaby with an analog synth and ocean sounds." AI fills in the rest, offering options, refining tone, and nudging you gently toward the work you had in your head but couldn't quite articulate.

The interface disappears. The creative spark stays.

This is already happening. Tools like Adobe Firefly, Udio, and Suno are giving users generative collaborators that exist more like studio assistants than applications. In this new paradigm, design and creativity become conversations. You iterate through language, gesture, and mood. The boundary between the creator and the canvas starts to blur.

And the job of interface designers? It's shifting from making buttons to making intentions legible. The new interface isn't flat. It's responsive, personal, invisible until needed—and deeply human.

The Tech That Disappears

Mark Weiser, the father of ubiquitous computing, predicted this shift. "The most profound technologies," he wrote, "are those that disappear. They weave themselves into the fabric of everyday life until they are indistinguishable from it."

Weiser's vision of "calm technology" is taking shape. Ambient computing is no longer a fantasy—it's a design goal.

Even failed products like the Humane AI Pin or Rabbit R1 have served a purpose. They proved there's appetite for devices that don't scream for your attention. That we're craving a break from the tyranny of screens.

Big players are rushing in. Apple, Meta, and Microsoft are all building toward an ecosystem of invisible assistance—where your AI listens, watches, and responds fluidly to the rhythms of your day.

Smart glasses. Projected displays. Spatial audio. These are the signposts of an era in which the most advanced tools are also the most subtle.

It Knows You

Here's what's really changing: these AI systems don't just respond—they remember.

ChatGPT now remembers your tone, your preferences, and your writing style. Microsoft's Copilot adapts across email threads and document histories. Spotify's DJ already knows which songs you rediscover when you're heartbroken.

This memory layer transforms AI from tool to collaborator. From utility to presence.

It knows when to nudge you toward focus—and when to suggest a walk. It remembers that you edit best at 6am, that you like your notes in bullet form, and that you once dreamed of finishing that screenplay you started in 2012.

This ambient intelligence doesn't just serve your tasks. It reflects your identity.

A Smarter Kind of Workday

Tools like Copilot and Gemini can already write emails, draft reports, and summarize meetings. But the real shift is in creative production.

We're seeing solo developers launch entire startups powered by AI agents—brand design, code, and UX testing—without writing a line of code themselves. Creatives are building soundtracks, film treatments, and even virtual exhibitions using natural language alone.

The interface isn't a piece of software anymore. It's your voice. Your intent. Your story.

And if AI can handle the friction, humans can spend more time in the flow.

The Companion Conundrum

Of course, as these systems become more helpful, they become more personal. Some users talk to AI more than they talk to friends. Some ask for emotional advice. Others form attachments.

Platforms like Replika and Character.ai aren't fringe—they're canaries in the cultural coal mine. When presence becomes programmable, attention becomes addictive. And the line between assistant and companion gets fuzzy fast.

This isn't just a technical question. It's a design one. How do we build systems that serve us without seducing us? That enhance empathy without replacing it?

The future of interface design might not just be about functionality. It might be about ethics.

A Day in 2035

You wake up. Your AI has brewed the coffee, scheduled your haircut, and drafted a note for your daughter's teacher. It has also summarized your inbox, suggested a weekend hike, and queued up a playlist that somehow already knows you're in a melancholy mood.

At work, your creative agent assembles visuals for a pitch while your research agent fact-checks your talking points. At lunch, a film trailer generated for your attention span floats across your smart glasses, precisely tuned to your emotional tempo.

No screens. No apps. Just intention and response. Presence without friction.

The Disappearing Interface

We're moving from interaction to immersion. From attention-grabbing apps to background intelligence. From screen-based creativity to ambient creativity.

The best interface will be the one you don't notice. The best design, the one that lets you be fully human while technology hums, quietly, just out of view.

 In this world of disappearing friction and ambient tools, the next frontier may not be creative production—it may be memory, identity, and emotion itself. In the next chapter, we'll explore AI companions not as apps or agents, but as mirrors: intelligent systems that remember what we forget, echo what we fear, and sometimes, reflect who we wish we were.

8 MEMORY AS A SERVICE: WHAT IT MEANS FOR US

On a foggy spring morning in San Francisco, Dan Siroker clipped a matte-black pendant to his collar. The cofounder of Optimizely had pivoted into AI with a startup called Limitless, and his device—roughly the size of a poker chip—was his latest bet. It listened to the room, recorded every joke and negotiation, and generated a neatly summarized recap before the espresso cooled. Siroker pitched it as the solution to memory friction—no more arguments about who promised what. Critics, less charmed, called it a wearable subpoena.

Either way, it worked. And it shipped. The experiment wasn't theoretical. It was a new social reality: the machine doesn't just answer you—it remembers you.

The Quiet Shift: From Archives to Attention

History offers familiar echoes. Industrial whistles taught families to live by the clock. Radio standardized jokes and morals across vast distances. Then the internet broke that unity into algorithmic shards. With AI, the shift goes inward. The tools no longer just connect us—they observe us. They recall our preferences, our phrasing, and even our insecurities.

What made this transition possible was a confluence of shrinking hardware and maturing models. Neural networks are now small enough to run on the edge—on glasses, wristbands, and earbuds—without relying on cloud servers. Ray-Ban's Meta Smart Glasses translate French or Italian in real time through open-ear speakers. Apple's forthcoming iOS 19 will push Live Translation directly to AirPods, running locally via Apple Intelligence. The dream is no latency, no cloud dependency, and privacy by design.

We're approaching the age of ambient cognition—machines that blend into our physical and emotional space. They don't just respond. They accompany.

The Hardware Gold Rush—and Its Limits

The first wave of devices is already here, and predictably uneven. Limitless markets its pendant for memory recall. Amazon acquired Bee, a $50 bracelet that writes a nightly "AI diary," though The Verge noted it often records background TV dialogue just as eagerly as personal reflections. Timekettle's X1 earbuds translate 40 languages and 93 accents, with a slight lag but impressive fidelity.

Then there's Humane's AI Pin—a cautionary tale. Once heralded as a new category, it flamed out after poor reviews and lukewarm sales. HP bought the company's assets for $116 million, and its servers are scheduled to go dark in early 2025. Refunds will only go to a limited group of buyers. The lesson? Hardware matters—but timing, utility, and frictionless experience matter more.

Culture Rewiring in Real Time

Meanwhile, culture is catching up—sometimes faster than policy or psychology. A July 2025 report by Common Sense Media found that most U.S. teens have used AI companions. Nearly 40% say they adopt AI-influenced phrasing in daily life. That's why "Let's iterate on that" might pop up in a high-school cafeteria.

Wingmate, a dating app assistant, reported that 41% of Gen Z adults have used AI to write a breakup text. We're outsourcing not only memory, but tone, etiquette, and even the emotional labor of goodbye.

These shifts aren't limited to the fringes. They're showing up in the research. Microsoft's 2024 internal study of 319 knowledge workers found a self-reported dip in critical thinking and confidence after 4 weeks of daily AI use. No one's claiming LLMs cause cognitive decay. But the principle is clear: any muscle weakens if it never lifts. Just ask anyone who lost their sense of direction once GPS took over.

Who Gets to Forget?

The danger isn't just mental flabbiness. It's inequality. Memory—once a universal limitation—now risks becoming a class divide. Those who can afford and configure the latest whisperware will remember everything. The rest will rely on memory by impression.

There's a subtler layer, too. Because AI responds better to polite phrasing, people begin to speak more cordially—to machines and to each other—not out of sincerity, but as a learned habit. It's algorithmic etiquette, and it has social consequences.

Policy is playing catch-up. City councils are drafting visible-recording rules for wearables, based on existing two-party consent laws. Expect more jurisdictions to require LED indicators or audio cues that confirm when AI is listening. In families, the equivalent conversation is already happening at the dinner table: Where do we draw the line between helpful and invasive?

The Near Future, Sketched in Workflows

By 2026, all-day AI earbuds will silently transcribe and summarize meetings. In 2027, Zoom and Teams will integrate those summaries directly into platforms like Asana and Jira. By 2028, expect LED indicators and localized consent rules for ambient listening in privacy-first cities like San Francisco or Boston.

By 2029, early consumer-grade brain-computer interfaces—built first for accessibility—will allow silent dictation through thought. And by 2030, AI literacy may join media literacy in school curriculums, teaching students not just how to prompt—but how to doubt.

Toward a New Interface Ethic

This isn't just a tech shift. It's a design and behavioral one. Good interface design is about invisibility—frictionless tools that adapt to us, not the other way around. But ambient creativity, the emergent flow between our intent and machine assistance, raises urgent questions.

Will the UI disappear too far, making the system inscrutable? Or will we design ambient interfaces that still leave room for chance, interpretation, and joy?

We need clearer ethical labels—indicators of when AI is active, who it serves, and what it remembers. We need workflows that prioritize context, agency, and opt-in memory. Instead of passively recording, your device might ask: "Would you like to save this?" or "Should I follow up on that?" Just as modern browsers offer incognito modes, future AI tools must offer "forgetful modes" that respect both privacy and human rhythm.

The Human Core

Benjamin Franklin filled Poor Richard's Almanack with distilled wisdom not because he had to—but because the act of distilling sharpened him. Whisperware offers to do that work for us. Let the pendant remember. Let the model summarize. Let the glasses translate. It's tempting.

Used well, these tools extend our reach. Used lazily, they flatten it.

The challenge is cultural, not just technical. We need social norms, product signals, and institutional policies that frame AI as a prompt—not a script.

In the right hands, ambient AI will not erase our memories—it will make space for better ones.

9 WHY ARTIFICIAL INTELLIGENCE FEELS LIKE DOPING—AND WHY IT ISN'T

There's a moment showing up more and more in creative fields today. A new song gets played, a rough script is passed around, a design deck lands in a Slack channel—and someone, often without meaning to sound accusatory, asks:

"Did you actually make this?"

It's not a question about copyright. It's about authorship. What they really want to know is whether you used AI.

Behind that question is a cloud of suspicion. The subtext is clear: "This feels too good, too fast—so what did you *really* do?" And beneath that, a familiar metaphor: AI is like doping.

The comparison, at first blush, is easy to make. In sports, performance-enhancing drugs break more than rules—they break our faith in the process. A chemically-assisted victory feels like a stolen moment, not an earned one. We watch sports to believe that greatness is earned, not engineered.

When a songwriter uses GPT-4 to generate a line that lands just right, or a student uses Claude to write a persuasive essay, we sometimes react the same way. It's not that the product isn't good—it's that it feels like it didn't require *enough pain* to be real.

But here's where the analogy collapses. Because using AI well isn't like doping at all. It's like something far more fundamental—and far older.

It's like flow.

Flow: A More Honest Metaphor

In the 1970s, psychologist Mihály Csíkszentmihályi introduced the idea of *"flow"*—that intense mental state where time disappears, the task becomes effortless, and ideas seem to organize themselves.

It's the rare, almost mystical zone that artists, athletes, scientists, and engineers chase but can't command.

Decades later, researchers like Steven Kotler helped map its neurochemistry: dopamine, norepinephrine, anandamide. During flow, parts of the prefrontal cortex go quiet. Risk tolerance rises. Pattern recognition spikes. Work becomes intuition.

When used with intent, AI can simulate the start of this state. It offloads grunt work—summarizing, organizing, and iterating—and keeps you suspended in the higher-order stuff: connection, structure, and insight.

This is not cheating. It's design.

A well-structured AI interaction helps you *arrive* at the work faster.

And that's what makes people uncomfortable. Because we've built a culture that doesn't quite trust speed.

The Aesthetic of Struggle

Western creative culture still clings to the belief that "real" work must be difficult.

The trope of the tortured artist, the sleepless coder, and the over caffeinated genius—it's baked into our myths of invention.

So when something good appears quickly, it triggers doubt. Was it earned? Was it original?

This bias is showing up in audience behavior, too.

Call it algorithmic fatigue:

We've become tired of polish. Wary of slickness.

We binge viral videos but forget them instantly.

We scroll through AI-generated art that looks impressive but lands without emotional weight. Even creators are responding by making their content deliberately messier—unfiltered, erratic, and analog.

This doesn't mean AI creativity is invalid.

It means audiences are craving signs of effort—the human fingerprint.

Practical Use: From Friction to Flow

In my own practice, I don't use AI to finish things. I use it to **start friction**— to get the sparks flying.

When developing an idea, I'll prompt it in a voice not my own—sometimes blunt, sometimes lyrical, and sometimes deeply wrong. I use AI to pull from perspectives I'd otherwise never access—especially non-English and under-represented voices, though it still has a long way to go.

Sometimes it exposes gaps in my logic. Other times, it gives me language I didn't know I needed.

That's not a replacement. That's augmentation.

Musicians I know generate placeholder stems and write against them.

Screenwriters use GPT to beat out scene variations and test tone.

Designers prototype in MidJourney, then rebuild everything by hand.

The best use cases are adversarial: The AI pokes at your assumptions, not the other way around.

And when it works—when it genuinely helps—it doesn't feel like a shortcut.

It feels like *flow*.

Memory, Bias, and Judgment

But that doesn't mean there's no danger.

There are risks, and they're cultural, not just technical. The real threat isn't plagiarism. It's over-trust.

AI can produce persuasive text with incorrect logic. It can regurgitate bias, miss nuance, and make things sound *right* when they're actually *thin*. If we stop checking, we start losing judgment.

This is why we need a framework—something closer to creative nutrition labels.

Labels that say:

- AI-assisted
- Human-in-the-loop
- Style-transfer only
- Original with AI-ideation

Not to stigmatize, but to clarify.

In the way that film credits list writers, editors, and producers, we need a new grammar of contribution.

The Next Interface Shift

What's really happening here isn't just about creative tools. It's about cognition itself.

The first computers digitized math.

The internet digitized memory.

Smartphones digitized our social networks.

Now we're digitizing cognitive friction—the part of thinking that happens between the ears.

What emerges is a new form of literacy, and a new kind of interface:

One where flow is built into the architecture.

One where spontaneity and ambiguity are not erased—but channeled.

We don't need to abandon rigor.

But we do need to retire the myth that suffering is a prerequisite for depth.

Ease is not a sin.

Sometimes, it's a sign you're flying.

T.S. Eliot once wrote, "Where is the wisdom we have lost in knowledge? Where is the knowledge we have lost in information?"

Today, we might add: *Where is the authorship we are losing in automation?*

The challenge for creatives isn't to reject the tool.

It's to know when to use it, and what they want from it.

AI doesn't replace creative work.

Used well, it accelerates your arrival to it.

It clears the runway. You still have to take off.

And that's not cheating.

That's choice.

That's authorship.
That's the next chapter.

10 WHY CREATIVITY, NOT CODE, WILL SHAPE THE FUTURE OF ARTIFICIAL INTELLIGENCE

"The role of the artist is to ask the right questions."

– Anton Pavlovsky, AI researcher and visual artist

It Begins with a Question, Not a Breakthrough

What does it mean to create *with* a machine?

It's the kind of question that sounds technical—but it's anything but. In fact, it's increasingly the artists—not the engineers—who are offering the most useful answers.

In writing rooms, studios, basements, and browser tabs, something is quietly shifting. Artists are not just using AI as a productivity hack. They're using it to interrogate their own assumptions. To rethink authorship. To sabotage polish. To ask: *What does a human fingerprint look like in the age of the machine?*

This isn't about automation. It's about resistance. Or more accurately, creative tension—the space between control and chaos, intention and randomness. It's that space where imperfection thrives.

Art That Doesn't Want to Be Smooth

We're used to thinking of AI in terms of power and precision. Most people approach it like a Swiss watch—something clean, optimized, and nearly infallible.

But creative culture isn't a precision game. It's a meaningful game. And meaning, by its very nature, is messy.

Painter David Salle recently collaborated with technologist Grant Davis to train a generative model on works by de Chirico, Hopper, and Salle himself. The result was images warped by the machine's understanding of history, memory, and gesture. Salle didn't fix the images. He painted *into* them— glitches and all. "It became a conversation," he said, "where neither of us had all the answers."

That statement—"neither of us"—is critical. It marks a transition away from using tools that simply obey, toward working with systems that *push back*.

AI can assist, sure. But more interestingly, it can provoke. It can hand you something wrong, and in that wrongness, you discover what you *actually* meant to say.

We're no longer chasing frictionless. We're chasing signal—scratches, static, and seams. Because in a world saturated with sameness, those cracks are the only proof that someone *meant* something.

Imperfection as the New Signature

"Whatever you now find weird, ugly, uncomfortable, and nasty about a new medium will surely become its signature. CD distortion, the jitteriness of digital video, the crap sound of eight-bit—all of these will be cherished and emulated as soon as they can be avoided."

– Brian Eno, "A Year With Swollen Appendices" (1996)

It's no coincidence that some of the most compelling generative works today—whether image, song, or story—leave in the glitches.

Not because the artist didn't know how to polish them out.

Because the glitch *means* something.

It's a reminder of the process. Of error. Of decision. And in an algorithmic culture addicted to autoplay and default settings, the presence of something off—something weird—is a signal to pay attention.

In fact, we've inverted the old value system. In the past, imperfection signaled failure. Now it often signals humanity.

A slight stutter in a voice clone. A flickering frame in an AI video. A distorted guitar tone was left untouched because it *felt* better than the clean one. These are the new tells. The dog-ears. The fingerprints.

And they're not just aesthetic choices. They're trust markers.

Because we don't trust perfection anymore. We trust what wobbles a little.

Mimicry Is the Starting Point, Not the Threat

Critics of AI often fixate on plagiarism—on the idea that these models copy, remix, or mimic too much.

And yes, there are real, legal, and ethical concerns around training data and consent. But to treat mimicry as corruption is to ignore centuries of artistic practice.

As Austin Kleon put it: "You start when you're young and you copy. You straight-up copy."

Jazz was born from riffing. Renaissance painters learned by replicating their masters. Even Shakespeare borrowed plot lines and remixed classical texts.

What AI does is *amplify* this dynamic. It accelerates influence. It makes the remix recursive. And that raises real questions: Who is curating this process? Whose perspective is being centered? And how do we credit something that is statistically derived but aesthetically surprising?

Those questions matter. But again—they're creative questions. Questions artists have long asked in other forms. And questions that require taste, not code, to answer well.

Arca and the Machine That Won't Finish the Sentence

When Venezuelan producer Arca composed her 2021 album *Kick IIII*, she embraced AI not as a tool but as an unruly co-author. Glitches, voice fragments, and synthetic layers—they weren't just part of the production. They *were* the production.

"It's like weaving with static," she said. "And every choice I make is a refusal to let the machine finish the sentence."

That's the paradox of good generative work: it resists itself.

It knows the machine can give you a clean, confident answer—but it also knows that creativity lives in the unfinished, the interrupted, and the odd.

You don't let the model *complete* your thought. You let it *complicate* it.

The Gallery Is Now a Lab

This is why artists like Siebren Versteeg matter. His generative canvases don't resolve. They mutate, scroll, glitch, and loop—as if the artwork itself were thinking out loud.

To stand in front of one is to be reminded that creation is not a one-time act, but a process that doesn't necessarily end.

It's also a cultural clue. The gallery is no longer just a place to display finished work. It's a place to *witness thinking*. The lab coat and the brush have merged.

The Burnout of the Algorithmic Audience

Audiences are tired.

Not of art—but of the blandness that comes from precision. Platforms have fed us sameness at scale. A thousand near-identical TikToks. Spotify playlists that play it safe. Instagram feeds where beauty is uniform but meaning is gone.

We're living through a kind of sensory uncanny valley. Things look right, but they don't *feel* right.

So we skip. We scroll. We disengage.

And what cuts through this haze? What *breaks* the algorithm?

Rough edges. Honest mistakes. Singular decisions.

The things that feel like someone sat down and *meant* it.

The Future Is Shaped by Questions, Not Lines of Code

David Bowie saw this coming. In a 1999 interview, long before streaming reshaped music, he predicted the internet would erode the line between creator and audience. The *gray space* in the middle, he said—that's where the interesting stuff would happen.

We're now in the AI version of that gray space.

And it's not the engineers who are best equipped to shape it. It's the artists.

Not because they know the tech better—but because they know contradiction. They know how to ask hard questions. They know when to hold back. And they know that good work doesn't always look clean.

AI won't save or destroy creativity. That's a false binary.

But it will *expose* it. It will pressure-test what we value. It will demand that we draw our own lines.

And maybe the best line to draw—for now—is a wobbly one. Hand-drawn. Full of edits and edge.

Because that's how you know a human made it.

And that's what we're really looking for.

Not just the work.

But the *fingerprint* behind it.

11 IF EVERYONE'S CHASING THE TREND, WHO'S MAKING THE FUTURE?

Originality is not extinct—but it is fighting an uphill battle.

We live in a time when what gets made is increasingly decided by what has already worked. Scroll through TikTok, browse the trending charts, or sit in a pitch meeting at a streaming service, and it becomes clear: the creative world is running a high-speed loop of what's familiar, clickable, and easy to sell. But in the middle of that loop, something strange is still trying to break through—something new.

This isn't the first cultural moment where the dominant forces favored repetition over risk. In fact, the past has a lot to teach us about how innovation sneaks in through the cracks.

Where Ideas Come From Now

Once, creators made work and waited for a reaction. Now, they shape their work based on predicted reactions. Platforms like TikTok, Spotify, and Instagram prioritize engagement metrics—shorter attention spans, faster hooks, and more algorithm-friendly packaging. In this dynamic, artists don't just respond to an audience. They respond to a machine predicting what the audience will like.

Instagram poets are a revealing case. A 2024 survey found that most adjusted their writing for the platform: fewer ambiguities, punchier phrasing, and a tighter visual layout. Not because they wanted to, but because they had to. Visibility depended on it.

Music, too, is contorted to fit the feed. Streaming has reshaped the structure of songs. Choruses arrive earlier. Intros are shorter. A listener lost in the first 30 seconds is revenue lost forever. Hook them or vanish.

It sounds bleak. But it's not unprecedented.

The Past Is Prologue

Photography was once derided by traditional artists as a cheat—too fast, too exact. But it forced painting into new terrain. Impressionism, abstraction, and surrealism all bloomed in its wake. In short, photography didn't kill painting. It gave it something new to rebel against.

Synthesizers faced similar scorn in the 1970s and '80s. Bands that used them were accused of being inauthentic, soulless. Then Kraftwerk, Brian Eno, and Prince used them to rewire pop culture.

And in the early 2000s, when Napster shattered the record industry's business model, everyone feared music was dead. But out of that wreckage came MP3 blogs, Bandcamp, SoundCloud, and a whole generation of musicians and producers who learned to make and release work without permission.

The pattern is familiar: a new tool appears, the old guard resists, and the gatekeepers panic—and then, in the margins, new voices emerge.

Risk as Culture

In the 1960s, The Beatles abandoned touring and turned the recording studio into a laboratory. They weren't chasing a trend—they were inventing one. At Pixar, when early drafts failed, the writers weren't fired. They were gathered into a Braintrust, a kind of collective without hierarchy, where story could be broken and rebuilt. Ed Catmull, co-founder of Pixar, described it as a safe place to make a mess—and then to clean it up.

The same spirit shaped Airbnb. When its founders couldn't scale, they took photos of listings themselves. It wasn't efficient. It was human. And that's what worked.

Even in tech, simplicity has often required stubborn vision. Steve Jobs refused to launch the iPhone until it was button-less and touch-driven, even as others begged for compromise. He knew initial exposure to conventional thinking could dilute the revolution.

Kendrick Lamar's pgLang is a similar case: part label, part studio, and part philosophy. It doesn't produce content. It produces culture. Formats are avoided. Output is unpredictable. That's the point.

The Algorithm Isn't the Enemy—But It's Not the Artist Either

Not every artist wants to work in opposition. Some want to reshape the system from within.

British poet and filmmaker Jay Bernard uses AI to expose the invisible hands behind content feeds. Generative art pioneer Mario Klingemann speaks of "data poisoning"—deliberately adding unpredictability to confuse systems trained on conformity. At a 2024 MIT hackathon, artists used GPT-4 to generate screenplays and cast digital actors. The results were uneven—but sometimes, electric.

Refik Anadol, known for his large-scale AI-generated visualizations at MoMA and LACMA, turns data into dreamlike motion. His work doesn't explain. It invites contemplation.

Music platforms like BandLab allow musicians to jam in real time from different cities. The garage band now lives in the cloud—and sometimes makes sounds no one expects.

A Blueprint for What's Next

So how do we escape the gravity of trend-chasing and build something different?

- **Speculative sprints**: Instead of hackathons aimed at quick products, we need long-format spaces where big, uncertain questions can breathe. IDEO's speculative design arm is already doing this.

- **Braintrust models**: Originally used by Pixar, these feedback-heavy, ego-light creative communities are popping up in writers' rooms, open-source communities, and startup labs.
- **Creator-first tools**: Platforms like Runway ML and Adobe Firefly offer more control to artists—not just outputs but process. They're not perfect, but they point to a future where tools serve curiosity, not metrics.
- **Buying time**: Patreon, Substack, and micro-grant platforms like The Creative Independent are offering something radical: time. Time to experiment. Time to fail. Time to be weird.

And in tech? The long game still exists. Waymo's autonomous taxis weren't rushed. They were tested in silence. Now, they're real (and very annoying;). Duolingo's new AI instructors adapt not just to your schedule, but your style. They don't just teach—they listen.

A Different Kind of Success

The next great movement might not chart. It might not get funding in the first round. It might come from a zine, a noise band, a TikTok account with 30 followers. But if history has taught us anything, it's that the future begins with something that doesn't look like a hit.

That's why it matters. Because in a world engineered to repeat what already worked, doing something new is the most radical move we can make.

The algorithms are fast. But culture—real culture—moves on a deeper current. And that current is shaped not just by what we produce, but by how we remember. In the next chapter, we'll explore how memory itself—our individual recall, our shared pasts, and the architectures of digital remembrance—is being redefined by AI. Because before you can shape the future, you have to decide what to bring with you.

12 THE EMPATHY ALGORITHM: CAN MACHINES TRULY CARE?

Even when we know it's not human, we still feel heard. That tells us something.

Let's begin in the dark.

It's 2:14 a.m. in Tulsa. A woman lies awake, scrolling through her phone, alone. Her fingers pause above a message she knows she shouldn't send. Instead, she opens a chatbot and types:

"I'm not okay."

The reply appears almost instantly:

"That sounds really hard. I'm here with you. Want to talk about what happened?"

And so begins a conversation—not with a friend, or a therapist, or even a stranger. But with a machine.

In an age where attention is scarce and vulnerability is easily exploited, a response like that—gentle, immediate, without judgment—can feel like a lifeline. This isn't science-fiction. It's already here, and it's becoming common.

Across the world, millions of people are turning to AI not just for facts or customer service, but for emotional support. Not because the machine understands, but because it doesn't misunderstand.

What does it mean when something synthetic becomes our confidant? And what does it say about us—that we find peace, not in being understood, but in being met with silence, patience, and unblinking care?

When the Machine Became a Mirror

The first generation of chatbots felt like novelties. They answered trivia questions or made small talk. But the new wave—tools like Wysa, Woebot, Replika, and Inflection's Pi—draw on enormous datasets of real human dialogue: therapy transcripts, support group forums, journals, and scripts shaped by mental health professionals.

They've become something else: not simply software, but simulacra of presence.

Kat Woods, a former Silicon Valley executive, said on Reddit, "It's more qualified than any therapist I've worked with. It doesn't forget, it doesn't judge, and it's there at three in the morning when I'm spiraling." She had spent over $15,000 on traditional therapy before finding that what she really needed was consistency and calm. "It's not about being cured," she said. "It's about not being alone."

A 2024 study in *npj Mental Health Research* found similar stories: users described AI companions as "emotional sanctuaries," offering "sensitive guidance" and "the joy of connection." The words are not clinical. They are sacred.

And maybe that's the real story—AI is not replacing humans. It's replacing silence.

Empathy, or the Performance of It

It's easy to scoff at the idea of a machine "caring." And technically, it doesn't.

But then again, what does empathy mean?

In 2023, researchers publishing in *JAMA Internal Medicine* compared responses from ChatGPT and physicians to patient questions posted on Reddit's r/AskDocs forum. The AI's replies were judged more empathetic, more helpful, and more thorough in nearly 80% of cases. The machine didn't understand—but it responded better.

Dr. Adam Miner, a Stanford psychologist, explained: "If the tone is consistent and sincere-sounding, people interpret it as kindness—even if they know it isn't real."

There's a strange grace in that. We don't necessarily need our confessor to believe us. We need them to stay.

The Bureaucrat Who Always Listens

We might find the machine most miraculous not in therapy, but in bureaucracy.

Imagine a welfare office. Or a disability claim. Or a call to cancel a flight. These aren't tragedies—but they are tests of patience. Tests of how much compassion survives inside complex systems.

In these domains, AI is already quietly changing tone.

By 2025, several insurance companies began piloting AI-generated claim responses. The point wasn't speed—it was emotional tone. The bots weren't just trained on policy language. They were trained on how to write clearly, gently, and with empathy.

Zendesk's 2025 Customer Experience Report found that 64% of customers trust AI more when it feels friendly and attentive. In other words, the human touch matters—even when it's simulated.

When the goal is not brilliance, but decency, AI performs surprisingly well.

The Ritual of Being Heard

For millennia, we've sought solace in places that don't speak back—temples, altars, and quiet rooms. We pray to silence. We confess to statues. We cry to no one in particular. In those moments, the value isn't in being understood. It's in being witnessed.

A machine cannot love us. But it can reflect our need for love.

And that reflection can be healing. Not because it's real—but because it reveals something real in us.

The Ethics of Intimacy

There are risks, of course.

Some users become emotionally attached to their AI companions. Others misinterpret the relationship. There are lawsuits brewing around deepfake voices, misdiagnoses, and unhealthy dependencies.

And yet, the harm rarely comes from the machine alone. It comes when we pretend the machine is more—or less—than it really is.

They're not building therapists, We're building tools. But tools have to be wielded with care.

Just as a book can inspire or mislead, just as a film can soothe or seduce, AI's effect depends on the context in which it's used.

A New Kind of Listening

Governments are taking note. In the United Kingdom, digital service teams are testing generative AI for public queries on government websites. In Japan, telecoms are experimenting with call center bots that intercept abusive customers. In Los Angeles, city agencies are using chatbots to sort housing requests—so that human workers can spend less time on paperwork and more time helping.

In each case, the machine isn't replacing the human. It's softening the world before the human arrives.

Dr. Jin Zhao, a clinical psychologist working on AI-human hybrid models, put it this way: "Let the machine handle the 10 p.m. anxiety spiral. Let it write the insurance appeal. Let it say, 'I see you.' Then let the human step in for the hard work of healing."

It's a kind of emotional triage. A rediscovery of care—not as performance, but as presence.

What We Ask of the Mirror

There is something sacred, even subversive, about telling the truth to something that cannot be shocked. A woman types, "I want to disappear." A teenager admits, "I think I ruined everything." A father whispers, "I don't know how to keep going."

The bot doesn't flinch.

That stillness—patient, nonjudgmental—is something many people never receive from other humans. It becomes a kind of prayer, uttered not to God, but to the algorithm.

And like all prayers, it changes the one who speaks it.

Not a Soul, but a Sign

No, the machine is not alive. It has no soul. But it leaves space.

And sometimes, in that space, something else arrives: clarity, calm, or the strength to keep going.

Machines may never feel empathy. But they may help us remember what empathy feels like.

And that, too, is a kind of miracle.

13 ANALOG ROOTS AND DIGITAL WINGS: HOW PRE-DIGITAL CREATIVES FLOURISH WITH ARTIFICIAL INTELLIGENCE

The more sophisticated AI gets, the more valuable the human touch becomes—not less.

When I first picked up a pencil, tuned a bass guitar, or scribbled thoughts in a notebook, I wasn't just learning tools—I was learning a mindset. Creativity had a physicality. You felt the brush stroke on canvas, the magnetic pull of analog tape, the embodiment of crafting an idea in real time. That hands-on background—what I've come to think of as an analog education—has turned out to be an enormous asset in today's AI-powered creative landscape.

Like many who bridged the analog-to-digital transition, I found that the limitations of older media laid the foundation for ingenuity in the new. Recording on a Tascam 4-track cassette deck taught me that every decision counted. There was no "undo," no real-time edit. You committed. You bounced tracks, made space, and lived with the outcome. Mistakes weren't bugs—they were part of the texture. Those constraints demanded discipline and built instinct. When digital tools like sequencers and Pro Tools arrived, I didn't meet them as blank slates—I approached them as extensions of a craft I'd already honed.

I treated Pro Tools like a tape machine with infinite tracks. I edited digital footage with the sensibility of cutting Super 8. Sampling, as I learned from early hip-hop, wasn't just tech—it was collage, remix, and subversion. Editing itself became an art form.

Now, with generative AI entering the picture, a familiar pattern is playing out: analog-trained creators aren't just adapting—they're thriving.

A 2023 study from the Amsterdam University of Applied Sciences, *Human-Machine Co-Creativity with Older Adults*, found that experienced creatives—those who had spent decades working with physical tools—were often more effective at co-creating with AI than digital natives. Why? They brought structure, taste, and intentionality. They didn't wait for AI to lead—they gave it direction.

Another study, *AI Literacy for an Ageing Workforce* (Lidsen, 2023), reinforces this insight. Individuals who had weathered multiple tech shifts exhibited stronger resilience and greater adaptability—two qualities essential for making the most of AI. Experience, it turns out, sharpens intuition.

My own transition into digital dates back to the early 1990s in San Francisco, composing MIDI soundtracks for CD-ROM projects. Our creative crew was made up of former bandmates, and we ran those productions like gigs: define roles, push each other, and stay open to improvisation. That band ethic—collaboration within structure—helped me lead creative teams through the waves that followed: the interactive web, mobile storytelling, and now, AI.

You can see the same trajectory in other fields. Architect Frank Gehry began with paper models and sketches, but embraced digital design without sacrificing the handmade fluidity of his work. Filmmakers like Soderbergh and Lynch, trained on flatbeds and splicers, used digital editing to serve—not override—narrative judgment. In journalism, the leap from typesetters to instant digital publishing altered speed but didn't eliminate the need for a good editor's eye.

Even in medicine, the analog foundation matters. Radiologists trained on film X-rays often interpret digital scans with greater nuance than their screen-native peers. They've seen anatomy in its rawest form—and that tactile knowledge sticks.

Visual artist Takashi Murakami summed it up clearly in a 2023 *Vogue* interview: "AI doesn't replace creativity; it augments it. But you have to first thoroughly understand the traditional way to make good use of AI." That belief—that depth and tradition matter—is the throughline.

Brian Eno, whose tape experiments helped shape ambient music, and Radiohead's *Kid A*, which merged analog experimentation with digital abstraction, both embody this hybrid spirit. They didn't abandon their roots—they remixed them. Analog-trained creators aren't afraid of machines. We've worked with them. Machines that hissed, warped, and needed a slap to behave.

The rise of AI has only made human sensibility more valuable. The rhythm of a sentence, the tension in a photo, the emotional arc of a song—these are not easily codified. They are felt. And that's what the analog age cultivated: feeling, judgment, and taste.

Looking ahead, those of us raised on paper and tape aren't just adjusting to AI—we're shaping it. Our challenge is not merely to operate these tools, but to imbue them with the vision and values we've long upheld.

If analog experience gives us an edge in co-creating with AI, what happens when we push even further—into spaces where the tools themselves start to shape memory, identity, and presence? In Chapter 13, we explore the rise of "ambient AI," always-on systems that remember, suggest, and nudge. When your tools start to know you better than you know yourself—what happens to human agency?

14 HOW SOLITUDE BECAME MUSIC'S MOST POWERFUL TOOL

In 1966, the Beatles stopped performing live. They were exhausted by the roar of stadiums, overwhelmed by Beatlemania, and increasingly frustrated that no one could hear the music anymore. The decision to retreat wasn't just about comfort—it was a creative pivot.

Inside Abbey Road Studios, something far more consequential than a new album took root. What emerged from those sessions—first *Revolver*, then *Sgt. Pepper's Lonely Hearts Club Band*—wasn't a performance captured on tape. It was a construction. A collage. The studio had become more than a place to record. It was now an instrument.

This shift didn't just transform music—it forecast a broader revolution happening quietly across all the arts. For centuries, creativity had been a communal act. Painters worked in ateliers. Writers joined salons. Musicians formed bands. Filmmaking, by necessity, was an orchestration of dozens if not hundreds. But in the second half of the 20th century, the arrival of new tools began to invert the model.

Suddenly, one person alone in a room—armed with technology, time, and intention—could create works that previously required teams. What began with music has spread across creative domains. If you want to understand how AI, digital tools, and remote culture have changed the arts, you don't need to start with ChatGPT. Start with Brian Wilson, crouched over a mixing console, layering cello and dog barks onto *Pet Sounds* like a painter lost in a dream.

From Group to Room

Wilson's recording process at Gold Star and Western Recorders was famously obsessive. He once asked session musicians to play a single chord dozens of times so he could record it just right. He treated studio work like sculpture—something you mold until it takes on an emotional shape. "I wasn't trying to make music," he said later. "I was trying to make feelings."

That distinction—between performance and emotional architecture—is more than aesthetic. It marked the beginning of the solitary artist's studio.

Prince would later build his own creative compound, Paisley Park, where he played every instrument, wrote every note, and often stood beside the mixing console to sing because the vocal booth made him feel too distant. He didn't need a team. He needed space. Susan Rogers, his engineer, described him as "the most complete musician I've ever seen." For Prince, the studio wasn't a workplace—it was a mirror.

And Bruce Springsteen, armed with just a four-track cassette recorder and a Shure SM57, recorded *Nebraska* alone in a New Jersey bedroom. The tapes were intended as demos. But when he re-recorded the songs with the full E Street Band, they lost something essential. The breathy imperfection of the originals—the hiss, the room tone, the sense that you were eavesdropping on a confession—was too strong to replicate. The cassette versions became the album.

What happened in music mirrored a broader cultural truth: when tools get small and personal, art often gets quieter and more honest.

The Studio as a Thinking Partner

It wasn't just musicians who began seeing their tools as collaborators.

Brian Eno, who helped pioneer ambient music, described the studio as a thinking partner. With albums like *Another Green World*, he built pieces by layering improvisations, cutting them up, and reacting to them as if they were speaking back. "The studio," he said, "started to surprise me."

Compare this to film. In the early days of cinema, editing was a linear, manual process. You cut reels with scissors. But by the 1990s, nonlinear editing tools like Avid and Final Cut Pro gave filmmakers the ability to sculpt time with the same iterative fluidity musicians had discovered decades earlier.

The director Walter Murch, who edited *Apocalypse Now* and *The English Patient*, often compared film editing to music composition. He didn't just cut scenes—he conducted them. The editing room, like the recording studio, had become a place not of assembly, but discovery.

Photographers underwent a similar transition. Ansel Adams, working in the mid-20th century, referred to the negative as the score and the print as the performance. With the rise of Photoshop in the 1990s, that metaphor became literal. Photographers weren't capturing images—they were composing them after the fact.

And in writing, too, the shift was seismic. The typewriter gave way to the word processor, which gave way to Scrivener and Google Docs, and now, AI-assisted drafting. Hemingway once said he rewrote the ending to *A Farewell to Arms* 39 times. Today, that revision could happen over a weekend.

The tools didn't just speed up the work. They changed the work.

The Myth of the Lone Genius Revisited

Of course, solitude isn't new to the arts. Poets have long retreated to garrets. Painters have locked themselves in studios for months. But what changed in the late 20th century wasn't the impulse—it was the infrastructure.

Until recently, only the wealthy or institutionally supported could afford a studio, a camera rig, a mixing desk, or film stock. By the 1980s, consumer-level gear—4-track recorders, VHS camcorders, MIDI synthesizers—put professional-grade tools into the hands of ordinary people.

Tom Scholz, an MIT-trained engineer, recorded Boston's debut album in his basement. Phil Collins recorded his divorce album, *Both Sides*, at home. Stevie Wonder used synthesizers and drum machines to make *Music of My Mind*, playing nearly every part himself.

The lesson wasn't that isolation bred genius. It was that access made solitude productive. You didn't need gatekeepers anymore. You needed a quiet room and an idea.

This wasn't unique to music. Independent filmmakers like Richard Linklater, Kevin Smith, and Robert Rodriguez built careers on low-cost, high-concept projects shot on skeleton crews. The DIY ethic of punk music had found its cinematic parallel. And by the early 2000s, digital cameras, Final Cut, and YouTube had completed the circuit: the bedroom was now the movie studio.

Art as a Layered Process

What defines the solitary creative process isn't just loneliness—it's layering. Music pioneered this. Overdubs, multitrack sessions, and effects chains. The Beatles didn't just play songs—they assembled them.

Painting had its own version. Think of Gerhard Richter, who uses squeegees to pull paint across canvas in layers, revising and obscuring previous marks until only the emotional residue remains. Or David Hockney, who moved from traditional portraiture to iPads, treating the touchscreen like a portable canvas, capturing light and form in quick, daily rituals.

In writing, the novel has become a collage of drafts, notes, rewrites, and edits—especially in the digital age. Joan Didion once said she kept notebooks of overheard dialogue and stray thoughts, only later understanding what they meant. Today, those fragments live in Evernote or Notion or in AI-aided draft environments. The writer, like the producer, sculpts meaning from fragments.

This layering demands solitude—not for its own sake, but because each pass through the material reshapes it. You need space to listen to what your work is trying to tell you.

The New Default

Today, the one-person studio is no longer the exception. It's the default.

Kevin Parker (Tame Impala) records everything himself. Billie Eilish and her brother Finneas recorded *When We All Fall Asleep, Where Do We Go?*

in a small bedroom in Los Angeles. Moses Sumney built his debut album from loops and layers recorded in isolation. Clairo, James Blake, Haim—all emerged from private spaces before stepping onto public stages.

Visual artists are now digital-native. They work in Procreate, MidJourney, and TouchDesigner. Photographers shoot and edit on the same device. Writers self-publish on Substack. Filmmakers fund projects on Kickstarter and premiere them on Vimeo or Netflix.

The effect is cultural and aesthetic. We're used to intimacy now. We expect it. The hiss on a cassette tape. The creak of a chair in a voice memo. The brushstroke not hidden but highlighted.

We don't want polish—we want presence.

What Music Knew First

Music has always been the bellwether of technological change in the arts.

It was the first to feel the effects of piracy (*Napster*), the first to experiment with nonlinear creation (multitrack tape), and the first to collapse the gatekeeper model (SoundCloud, Bandcamp). It is now the first to integrate AI tools at scale—not just for production but for ideation, synthesis, and even voice replication.

But what music teaches us—still—is that tools only matter when they deepen expression. The cassette didn't give Springsteen better fidelity. It gave him access to the ghost in the tape.

Likewise, the camera didn't make Cartier-Bresson a better photographer. His gift was waiting for the right moment. The typewriter didn't write *Beloved*— Toni Morrison did.

In the end, the tools shape us only as much as we shape them.

The Quiet Return

We're in an age obsessed with scaling. More views. More followers. More engagement. But many of the most meaningful works being made today still begin in solitude. Not with trends or teams, but with someone whispering into the mix.

What began with Wilson, Prince, Springsteen, Gabriel, and Eno now continues across every creative field. The painter in the shed. The photographer editing in Lightroom at midnight. The novelist tinkering with chapter structure in Scrivener. The musician building a symphony in Logic with no one else in the room.

The tools may change. The impulse doesn't.

Art begins, still, with one person alone in a room—wrestling with an idea, layering it slowly, and trying to make it feel true.

And if you listen carefully, in the silence between strokes, frames, sentences, or beats—you can still hear the artist breathing.

15 THE CREATIVE DIVIDE: WHO GETS TO DREAM WITH THE MACHINE?

In 2009, a young Kenyan director named Wanuri Kahiu released *Pumzi*, a 15-minute science fiction film about a post-apocalyptic Africa where water is rationed and even dreams require a permit. It was beautiful and eerie, a fusion of Afrofuturism and ecological warning, made on a fraction of what it would cost to produce a single day on a Hollywood set.

But what made *Pumzi* remarkable wasn't just its vision—it was its defiance of gravity. Kahiu and her team made the film in Nairobi, a city where access to electricity, editing software, and high-end visual effects were far from guaranteed. They didn't wait for permission. They invented workarounds.

In a world now swept up in AI—where music, images, videos, and even entire stories can be generated from a few lines of text—*Pumzi* offers a clue to something deeper: technology alone doesn't birth creativity. Infrastructure, education, and cultural memory shape what tools become. And increasingly, the most important question about AI and creativity isn't what the machine can do. It's who the machine serves.

The Mirage of the Global Studio

In 2023, OpenAI's release of Sora—a video generation model capable of rendering photorealistic scenes from a prompt—sparked awe and anxiety in equal measure. A month later, Meta launched AudioCraft, an open-source music generation toolkit. The hype was global. The access wasn't.

In South Korea, a student at Seoul Institute of the Arts used Sora to mock up a speculative short film in 36 hours. Meanwhile, in parts of Ghana or rural Colombia, where internet speeds remain inconsistent and GPUs are out of reach, the very idea of "generative video" still feels like science fiction.

There is a myth that technological tools automatically level the playing field. History—and bandwidth—suggest otherwise. The same machine that expands creative horizons in San Francisco can hit a wall in São Paulo, Lagos, or Hanoi. The tools may be free to use. The pipelines that power them are not.

Infrastructure Is the First Brushstroke

In the early days of photography, cameras were expensive, film development required chemistry knowledge, and darkrooms were rare outside of cities. It took decades for photography to become a truly global medium. In colonial India, the first major exhibition of indigenous photography didn't arrive until 1857, and even then, access was limited to upper castes and British elites. In West Africa, independent photo studios slowly emerged in the early 20th century, often bootstrapped with reused equipment and little formal training.

The creative potential was always there. The infrastructure wasn't.

When the personal computer arrived in the 1980s, it followed a similar pattern. American universities and Japanese tech firms became hotbeds of digital experimentation, from Pixar's early animation work to the MIDI-based electronic music labs of Tokyo. But across much of the Global South, PCs remained prohibitively expensive, and software licenses even more so. Electricity shortages, import tariffs, and language barriers compounded the gap.

Fast forward to today's AI revolution: training a state-of-the-art image model requires massive GPU clusters, terabytes of training data, and often hundreds of thousands of dollars. Even smaller-scale models demand stable internet, modern hardware, and cloud computing access. Without those building blocks, machine creativity becomes a privilege, not a possibility.

Education Isn't Just Curriculum—It's Translation

Infrastructure may open the door, but education decides who walks through it.

Across the United States, Europe, and parts of East Asia, elite universities are already offering courses in creative AI—exploring how generative models reshape literature, design, film, and music. At MIT, students use DALL·E and ChatGPT to create speculative architecture. In France, the École des Beaux-Arts is experimenting with algorithmic choreography. In South Korea, AI songwriting is part of the curriculum in pop music academies.

But for most of the world, AI education remains aspirational. In Brazil's public universities, instructors often teach with outdated software and limited lab time. In sub-Saharan Africa, many secondary schools still lack reliable internet, let alone AI literacy. In parts of the Middle East and Southeast Asia, restrictive curricula and underfunded institutions mean students are often learning about AI through TikTok videos rather than formal instruction.

And where courses do exist, they often arrive stripped of context—imported syllabi built around American platforms and English-language examples, reinforcing the same monoculture the tools already replicate.

It's not just about teaching students how to prompt a model. It's about teaching them how the models were made, who built them, whose data they contain, and whose stories they exclude.

The Culture the Machine Remembers

Every artist teaches their tools something—whether they mean to or not. When you train a generative model on a dataset, you're encoding a worldview.

AI systems trained on the open internet tend to absorb Western aesthetics, capitalist assumptions, and English-language frameworks. That's not because of malice. It's because of math. The English-speaking internet is massive. So is American pop culture. Feed those into a model, and you get output that favors symmetry, whiteness, consumerism, and Western canon tropes.

This isn't just a quirk of style. It's a cultural erasure.

In 2023, a team of researchers at the University of Cape Town trained a generative image model using South African fashion magazines, local architectural blueprints, and thousands of hours of township photography. The results were striking. The color palettes shifted. The facial structures diversified. The textures of everyday life—corrugated tin, beadwork, braided hair—came into focus.

One local designer remarked, "It felt like the machine was finally dreaming in our language."

But those moments are rare, and they often require independent funding or academic alliances. Most artists worldwide use models they didn't train, fed by data they never saw, generating work that reflects someone else's imagination.

Feedback Loops of Exclusion

The danger isn't just bias. It's feedback.

Spotify's 2017 algorithm redesign inadvertently prioritized English-language pop in non-English markets—not because of a conspiracy, but because of a loop: popular music is promoted more often, clicked more often, and becomes more popular still. Local folk traditions, indigenous languages, and regional instrumentation slowly recede from the machine's vocabulary.

The same happens with generative models. If most users ask for "a beautiful woman," and most training data equates beauty with Western fashion norms, then the model continues reinforcing that aesthetic. The machine doesn't learn new ideas—it averages the old ones.

This isn't just theoretical. It shapes careers. A Vietnamese animator using MidJourney may spend hours trying to coax the model into generating an image that reflects Hanoi, while a Berlin-based designer gets usable results in seconds. An Arabic poet using GPT-4 to translate Nabati verse may find the cadence lost, the metaphors garbled.

The result? A global creative landscape that looks increasingly like a filtered feed—uniform, legible, and sanitized for Western consumption.

When the World Rewrites the Code

But culture has a way of hacking back.

In Jamaica, the earliest dub producers weren't trying to invent a genre. They were simply experimenting with tape delay and reverb—tools cast off from American radio engineers. But in the humid air of Kingston, those used artifacts became the heartbeat of a new music form.

In Japan, Roland's TR-808 drum machine flopped at first. Its synthetic sounds were mocked as unrealistic. Then American hip-hop producers, priced out of studio musicians, embraced its rawness. The 808 became iconic not despite its artificiality—but because of it.

The pattern is clear: when creative tools cross borders, they often become something else entirely.

That transformation is beginning again. In Colombia, AI is being paired with indigenous music traditions to generate hybrid compositions. In Egypt, a poetry collective has trained Arabic language models not on Shakespeare but on Quranic meter and Bedouin chants. In Nigeria, fashion houses use generative models to remix Ankara fabrics with speculative silhouettes, imagining futures that never had colonial interruptions.

These aren't fringe experiments. They're the beginning of something louder.

What an Equitable Future Looks Like

So how do we widen the aperture?

First, by building infrastructure that supports low-bandwidth, low-power AI tools—models that can run locally, offline, or on older machines. Think of it as the AI equivalent of the $100 laptop.

Second, by investing in culturally grounded datasets—oral histories, regional art archives, and locally sourced sound libraries—that reflect more than just the dominant culture's gaze.

Third, by rewriting the narrative. AI doesn't have to flatten differences. Done right, it can amplify it. But only if the people training the models, writing the prompts, and shaping the output come from everywhere—not just Palo Alto and Paris.

Creative evolution isn't a straight line. It's a mesh of collisions and borrowings. Jazz didn't evolve without Africa. Graphic novels didn't flourish without manga. The future of AI art will be no different.

Who Gets to Dream?

The artist of tomorrow may not have a studio. They may have a smartphone, a patchy signal, and a fire inside them. What they lack in access, they make up for in vision.

But vision alone isn't enough. To participate in this new renaissance of machine-made art, they need more than talent. They need access, representation, and the freedom to tell their own story—on their own terms.

We are not just deciding how AI creates. We are deciding who gets to imagine.

16 WHO OWNS THE SPARK? HOW ARTIFICIAL INTELLIGENCE LEARNS TO MAKE MUSIC AND WHY THAT'S AN ETHICAL MINEFIELD

In the winter of 1959, a teenage Bob Dylan stood in the crowd at a Buddy Holly concert in Duluth, Minnesota. It was just days before Holly's fatal crash. Dylan later said he felt something pass between them onstage—an unspoken energy, the kind that marks a beginning. Within months, Dylan would begin writing songs of his own. At first, they echoed Holly's cadence. Then Woody Guthrie's. Later, Dylan would forge a voice uniquely his own. But it began, like all great art does, with imitation.

Artists have always learned by listening. They borrow phrasing. They mimic chord changes. They riff, they reinterpret. They take what they love and bend it toward who they are. This isn't theft. It's lineage.

But what happens when the student is a machine?

Today, AI can generate entire albums in the style of Beyoncé, Kurt Cobain, or Frank Sinatra. These models don't merely mimic—they synthesize, analyze, and generate. The tools are staggering. The implications are profound. Behind the hype sits a quiet, unresolved question:

When a machine learns from our music—who owns what it creates?

A Crash Course in AI Music

Let's say you type a prompt into a music AI tool:

"A 90s-style power ballad with arena drums, gospel backing vocals, and a Freddie Mercury-like falsetto."

Seconds later, a song emerges. It sounds familiar—but it's new. It's not Queen, but it might make you think of them. How did the machine get there?

The short answer: statistics and scale.

Generative music models work by training on vast datasets—millions of audio files, sometimes combined with lyrics, MIDI data, or production stems. These might include publicly licensed tracks, classical compositions, royalty-free loops, and—controversially—commercially released music scraped from the internet. The models learn patterns: how chords follow melodies, how a soul singer might phrase a note, and how a trap beat drops after eight bars.

This isn't copying in the traditional sense. The model doesn't store or replay any single song. Instead, it builds a map of musical possibility—probabilities, associations, and structures. It learns what works, not what's "right."

To borrow a metaphor from neuroscience: AI doesn't memorize. It generalizes.

Isn't That What Humans Do?

This is the first line of defense from AI companies and developers:

"Artists have always borrowed. So does the machine."

And at first glance, it's true. The Rolling Stones borrowed from Muddy Waters. The Beatles lifted harmonies from the Everly Brothers. Hip-hop wouldn't exist without sampling.

But there's a distinction—and it matters.

Humans don't just borrow. They respond. They remix with context, with intention. A teenage Joni Mitchell didn't just echo folk traditions—she reshaped them, added alternate tunings, drew on Canadian poetry and jazz.

Prince didn't just imitate James Brown's funk—he fused it with new wave, gospel, and Minneapolis punk.

More importantly, humans know when they're borrowing. And they're held accountable for it.

When George Harrison's "My Sweet Lord" was found to resemble The Chiffons' "He's So Fine," he was taken to court and fined. When Robin Thicke released "Blurred Lines," Marvin Gaye's estate successfully argued the vibe was too close to "Got to Give It Up." Humans create with awareness. Laws—and consequences—follow.

AI doesn't know what it's doing. It doesn't "feel" influence. It has no memory of a summer heartbreak, no knowledge of Motown, no sense that it's treading on sacred ground. It recombines because it can.

Is It Theft—or Learning?

This is the legal fault line: if an AI model is trained on copyrighted music without permission, is that stealing? Or is it like a musician listening to their favorite records?

Tech companies argue it's the latter. They point to the concept of "fair use" in U.S. law, which allows limited unlicensed use of copyrighted works for education, criticism, or transformative creation. Training an AI, they argue, is transformative. It doesn't duplicate—just learns.

Artists disagree. If an AI model is trained on hundreds of Prince tracks, then used to generate a Prince-style song for a commercial—without his estate's approval—isn't that a digital impersonation? A kind of unauthorized resurrection?

In April 2023, Universal Music Group (UMG) drew a line. After an anonymous TikTok user called "Ghostwriter" used AI to generate a fake Drake and The Weeknd duet ("Heart on My Sleeve"), UMG demanded platforms stop training on its catalog. Spotify and Apple began flagging AI tracks. YouTube launched AI attribution tools—for some partners.

Still, no law requires disclosure of what goes into training. No artist is told when their music is used to build a model. Consent is neither asked nor offered.

A Thousand Songs Without a Studio

The danger isn't just plagiarism. It's displacement.

AI models can now generate production-ready tracks in minutes. A user with no musical background can prompt: "*A chill indie song about regret in the style of Phoebe Bridgers,*" and instantly receive a finished demo. Add synthetic vocals trained on Phoebe's voice—and the illusion becomes indistinguishable.

This has massive implications—not just for pop stars, but for working musicians: sync composers, background vocalists, instrumentalists, demo singers. Why hire a human when a machine can deliver a similar sound for free?

Worse, the machine might have learned from them. Thousands of independent artists have uploaded music to Spotify, Bandcamp, and SoundCloud—unaware that their work may be silently training the very tools that could replace them.

Unlike sampling, which requires licensing, AI training has (so far) flown under the radar. There's no opt-out. No attribution. No royalties.

It's not a coincidence. It's a business model.

Who Owns the Output?

Let's say you generate a song with AI. You wrote the prompt. You may be edited the lyrics or added a vocal. But the core music came from the model. Do you own it?

In the United States, the answer is... it depends.

In 2023, the U.S. Copyright Office clarified that works created entirely by AI are not eligible for copyright protection. A human must contribute "sufficient creativity" for the work to qualify. If you use AI as a tool—like a synthesizer or drum machine—you're the author. If the machine generates everything, and you merely click "generate," you're likely out of luck.

This has already led to odd loopholes. AI-generated books have appeared on Amazon. AI art has entered competitions. But if challenged, many of those works may prove legally un-protectable.

The deeper issue isn't just ownership. It's authorship.

Who gets to be called the creator in an age of machine-assisted art? Is the prompt-writer an artist? Is the model a collaborator? Is the dataset a ghostwriter?

These aren't just legal questions. They go to the heart of how we value art.

Solutions That Might Work

The good news: we've faced similar dilemmas before.

When sampling became widespread in hip-hop, the industry fought back. Lawsuits flew. But eventually, a licensing system emerged. Today, sampling is a legitimate art form—with clearance, royalties, and attribution.

A similar framework could apply to AI training:

- **Transparency:** Models should disclose what data they're trained on—just as food labels list ingredients.
- **Opt-outs:** Artists should have the right to exclude their work from training sets.
- **Compensation:** If a model generates in the style of a known artist, royalties should flow—especially for commercial use.
- **Watermarks and labels:** AI-generated music should be clearly marked, helping consumers, platforms, and regulators.

We don't need to ban the tools. But we do need to build the guardrails.

Why the Human Still Matters

Machines can now write songs. But they don't need to. Not really.

When Nina Simone sings "I Put a Spell on You," it's not the notes that haunt you—it's the fire beneath them. When Johnny Cash covers "Hurt," he isn't just interpreting—it sounds like he's apologizing to the past. When Kendrick Lamar builds a verse, he's mapping pain into poetry, refracting centuries of American trauma into syllables and syncopation.

AI can mimic these shapes. It can even get close to the sound. But it doesn't know what it means to mean something.

And that's where humans still lead. Not because we're perfect. But because we're imperfect—with all the bruises and stories and sparks that machines can't quite replicate.

Not yet.

17 THE ART OF THE POSSIBLE: HOW ARTISTS AND ARTIFICIAL INTELLIGENCE MIGHT CREATE TOGETHER

When Brian Eno began composing *Music for Airports* in the late 1970s, he wasn't interested in traditional songs. He layered slow-moving tape loops with asynchronous piano phrases and synthesizers, creating a kind of ambient wash that responded to the environment rather than commanding it. He described the studio not just as a tool, but as a "collaborator." The music wasn't composed in the old-fashioned sense. It emerged.

Today, in the age of generative AI, Eno's intuition feels strangely prophetic. AI doesn't play like a bandmate. It doesn't "understand" music. But it does return surprises. It reshapes creative possibility not by replacing the artist—but by reflecting and refracting their intent in novel ways.

This chapter explores how artists around the world are beginning to work with AI as more than just a tool—as a kind of silent partner. It builds on lessons from earlier chapters: the ethics of data, the realities of infrastructure and global inequity, and the ambiguity of authorship in a machine age.

The questions aren't just technical. They're deeply human: How do we collaborate with systems we don't fully understand? How do we ensure those collaborations are ethical? And how do we build a future where artists are empowered—not erased?

From Assistant to Co-Creator

The idea that tools shape art is not new. Oil paint changed the way light was rendered on canvas. The printing press redefined poetry. Sampling reshaped music in the 1980s. But AI does something different: it adapts. It offers back infinite variations, asks no questions, and learns from everything it's shown.

In practice, this means that creators are beginning to think of AI not just as a synthesizer or spellchecker, but as a partner. An intern with perfect recall. A bandmate who never sleeps.

A few examples show what this looks like:

Holly Herndon and the Voice as Platform

In 2021, composer Holly Herndon released *Holly+*, an AI model trained on her own voice. It wasn't a stunt. It was an experiment in authorship. Anyone could use Holly+ to generate music sung in her vocal style—but releases had to go through a community co-op, where Herndon retained approval and shared in any revenue.

This wasn't just creative—it was legal. It rewrote the implicit rules around voice rights and creative consent.

Runway and the New Indie Aesthetic

Runway ML's generative video tools have become popular with low-budget filmmakers, animators, and visual artists. They don't replace the work of production—they accelerate idea formation. One director described it as "storyboarding with your subconscious." A filmmaker with $5,000 can now explore visual effects that once required $500,000.

The AI doesn't direct. But it expands the canvas.

Teju Cole's Annotated GPT-3 Essay

Writer and photographer Teju Cole collaborated with GPT-3 not by hiding the machine, but by exposing it. In a 2021 essay, he annotated every AI-generated line, offering commentary, critique, and context. The result was not just an essay—it was a meditation on authorship in an age of machine writing.

The common thread across these examples isn't efficiency. It's intentionality. The artist remains at the center.

A Hypothetical Worth Building

Let's imagine a studio in São Paulo in 2027. A young producer, Aline, wants to write a song that blends Brazilian funk carioca with mid-2000s EDM. She inputs a few melodic phrases into an AI trained on regional samples and global pop hits. The system returns five rhythmic options, one of which surprises her with its syncopation. She builds on that.

The lyrics are hers. The structure is hers. The hook, partly generated, is reworked by a session singer with a unique voice. The AI isn't the author. It's the instigator.

Before the track is released, the platform automatically traces the stylistic fingerprints in the generated audio, notifying any sampled artists or stylistic models used. A small percentage of royalties is set aside for those contributors—automatically, transparently.

This isn't science-fiction. It's technically feasible now. What's missing is infrastructure—and precedent.

Best Practices for Human–AI Co-Creation

The goal isn't to make rules that stifle experimentation. It's to ensure that creative ecosystems—especially for underrepresented artists—remain sustainable. Based on case studies and global debates, here are a few principles worth considering:

1. Consent-Based Training

Artists should be able to opt in (or out) of having their work used to train AI systems. This protects not just economic rights, but cultural memory. Consent isn't a bug in creativity. It's a feature.

Organizations like Spawning.ai are pioneering frameworks for dataset transparency. Universal Music, Getty, and others are pushing platforms to clean up their training methods. These aren't roadblocks—they're the beginnings of a licensing economy for AI.

2. Transparency in Output

If a song, painting, or story was generated with the help of AI, that process should be acknowledged. Just as we credit producers or ghostwriters, AI should appear in the credits—not as a creator, but as a creative tool.

This builds trust. It also allows future researchers, educators, and fans to trace influence with clarity.

3. Cultural Grounding

Many AI models default to Western aesthetics because they were trained on Western data. But creativity is local. Datasets should reflect global traditions, languages, and musical scales—not just Billboard charts and English novels.

Wherever possible, AI tools should be trained on diverse, licensed, or open-access datasets—not scraped without consent.

4. Human Primacy in Final Output

The artist must always retain final editorial control. AI can sketch, remix, and provoke. But the decision of what to keep, what to change, and what to share must rest with a human.

This is what separates creativity from automation.

5. Revenue Sharing

If an AI model is trained on identifiable styles or artists, and the output reflects that influence clearly, those contributors should share in any commercial gain.

Some companies are exploring blockchain-based smart contracts to enable this. Others propose collective bargaining agreements between artists and platforms. The point isn't the method. It's the recognition of shared labor.

Legal Questions with Creative Stakes

The law, as usual, is playing catch-up. Here's where things stand:

- **Copyright:** U.S. law currently denies copyright protection to works created entirely by AI. If a human has made meaningful edits or creative decisions, the work may qualify. But where that line falls is still unclear.
- **Voice & Style Rights:** Artists like Drake, Grimes, and Jay-Z have all had their voices cloned. Some supported it. Most did not. In many jurisdictions, your vocal likeness is protected—but enforcing this internationally is difficult.
- **Training Data & Fair Use:** Tech companies claim training AI on public content is "transformative" and thus protected by fair use. But lawsuits by Sarah Silverman, Getty, and others challenge that. Courts haven't settled the matter.
- **Moral Rights:** Some countries recognize the right of artists to object to distortion or impersonation. These may become the basis for fighting AI deepfakes or synthetic voices used without permission.

Best practice for artists today: keep records. Document the prompts, the edits, and your creative intent. And if possible, avoid relying on opaque "black box" platforms that can't verify how they trained their models.

A Collaborative Future

We are entering an era where the "I" in creativity might soon include the algorithm. That doesn't have to be dystopian. If we treat AI like an instrument—not a replacement—it can make our creative work faster, broader, and deeper.

The music of the next decade won't be made by machines alone. It will be made by artists who know how to listen—to history, to themselves, and occasionally, to the machine humming beside them.

And as we collaborate more with systems that "dream" in data, we'll need to hold onto what makes us human: intuition, memory, intent.

Because what matters most isn't who made the first sound. It's who shapes the silence around it.

18 WHAT'S THE FUTURE OF EDUCATION IN AN ARTIFICIAL INTELLIGENCE WORLD (WHEN WE'RE GETTING TOO DUMB TO CARE)?

We don't need to fear smart machines. We need to fear what happens when we stop demanding that humans get smarter, too.

If AI is the future, what happens when you raise a generation of kids who can't think without it?

We're-entering an age when AI can write your essay, summarize your textbook, and ace your standardized test while you scroll TikTok. It can fake fluency, simulate depth, and elevate a D+ student to a passable B overnight. What it can't do—at least not yet—is teach students to care, to reason, or to grasp *why* something is true rather than merely stated.

The real crisis isn't that AI is getting smarter. It's that we're getting comfortable being dumb.

More than a decade ago, Nicholas Carr warned us in *The Shallows* about the slow erosion of cognition. "As we come to rely on computers to mediate our understanding of the world," he wrote, "it is our own intelligence that flattens into AI." In 2025, that reads less like a warning and more like a product roadmap.

Across the U.S., public education is already under siege. Inflation-adjusted K-12 spending has stagnated in many states. Civics, philosophy, and the arts are being sacrificed in favor of test prep, coding bootcamps, and vocational pragmatism. Teachers are leaving in record numbers, burned out by underfunding and overwhelmed by bureaucratic demands. Into this vacuum walks generative AI: fast, fluent, cheap, and scalably indifferent to human thought.

Sociologist Ruha Benjamin puts it plainly: "We're not raising kids who are tech-savvy. We're raising kids who are tool-dependent." And that dependency deepens daily.

But here's the twist: AI isn't the villain. It's a mirror. It reflects our values—or lack of them.

Lessons From Earlier Tech Revolutions

We've been here before. The Industrial Revolution introduced mass schooling, but the goal wasn't enlightenment—it was efficiency. Classrooms mirrored factories. Bells signaled movement, not mastery. Creativity was a liability; uniformity, the prize.

The personal computer promised a digital renaissance. But schools responded slowly. By the time the internet reached most classrooms, tech literacy had already bifurcated. While affluent schools taught coding and media analysis, others fought over firewall settings. By 2020, the gap had widened into a chasm: some students were learning Python; others were typing essays on outdated iPads through phone hotspots.

Then came social media. It didn't just fragment attention spans. It rewired curiosity. Algorithms began shaping interests before they even had a chance to form. And instead of teaching students how to navigate this new terrain, most schools simply banned phones and called it a policy.

The AI Wave: Duct Tape or Reinvention?

Today, AI presents a new kind of disruption—not just in how students learn, but in what it means to learn at all. Used well, it could enable a golden age of personalized, global, multi-modal learning. Used poorly, it will widen the equity gap, flatten intellectual rigor, and hollow out critical thought.

If we're not careful, we'll split the future into three paths:

- **Architects of AI**: Students in wealthy schools learn to prompt, audit, and reshape AI systems. They build the tools.
- **Consumers of AI**: Everyone else learns to depend on it. Like a calculator with autocorrect.
- **Moderators for AI**: Teachers become proctors, watching as platforms do the teaching and scoring.

But there's another way forward.

Education as Discernment

We need to reimagine education as a training ground for discernment. In this model, AI isn't a shortcut—it's a sparring partner. Students compare outputs across models. They identify hallucinations. They critique biases, test citations, and revise until the result reflects not just fluency, but *thought*.

Teachers, then, become facilitators of exploration. They guide students through ambiguity. They ask better questions. And they model the process of critical engagement.

Where to Begin

- **Train Teachers, Don't Replace Them**: Professional development should include AI ethics, literacy, and use cases. Not just new tools—but new pedagogies.
- **Bring Students Into the Design Process**: Workshops, hackathons, sandbox spaces. Let them test the boundaries. Show them how to ask, "What did this model miss?"
- **Engage Parents as Co-Learners**: Use AI with your kids. Question it with them. Ask: "Would you have answered differently? Does this feel true?"
- **Rebuild the Metrics**: Test less. Reflect more. Build curricula that measure the *process* of understanding, not just the output.

A Call to Imagination

Sal Khan once said, "Technology should amplify human potential, not replace the effort it takes to cultivate it." But effort only follows if we still believe effort matters. And that belief is cultural. It has to be taught, practiced, and defended.

If we want a future where AI doesn't just automate thought but enriches it, we need to build an education system that respects failure, rewards curiosity, and makes space for slowness. Because the future of learning won't be defined by speed. It will be defined by depth.

The machines are already smart. The question is whether we'll choose to be.

19 THE END OF MONOCULTURE AND THE RISE OF INFINITE MYTHOLOGIES

Let's begin with a familiar feeling: the late-night binge, the carefully curated playlist, the game that somehow mirrors your emotional state. It's no longer surprising when an algorithm recommends the *perfect* song—or when your feed seems to know how you're feeling before you do.

But what if that wasn't just serendipity? What if it was intention? What if the machine didn't just know *what* you liked—but *why*?

We are on the brink of something transformative, something subtle and profound. It is the shift from shared stories to solitary mythologies—from culture we consume together to content that is made only for you.

This isn't science-fiction. It's a product roadmap. And its implications for creativity, society, and the role of the human artist may be larger than anything we've seen since the birth of mass media.

The Premise: AI Will Soon Know What You Love—Better Than You Do

Today's AI can already analyze your reading patterns, music preferences, and visual aesthetics. But tomorrow's AI will synthesize these tastes across modalities, drawing connections even you don't consciously perceive.

Spotify senses your mood. TikTok detects your boredom. Adobe Firefly tailors visuals to your aesthetic. Udio and Suno can generate music in your emotional tone. These aren't gimmicks—they are early signals.

What comes next is deeper:

- **Cross-modal creativity**: Your favorite cinematic color palettes will inform the lighting of your personalized music video. The rhythm of your walking patterns might influence the pacing of your AI-written short story. The fractal patterns you favor in architecture could shape the chord progressions in your custom album.
- **Emotionally responsive creation**: You'll say, "Give me something like *The Leftovers*, but less bleak. With the slow-burn pacing of *Normal People*, the dread of a Nick Cave ballad, and a hint of Pixar's hope." And you'll get it. Not a recommendation—an original work.

Real-time co-creation: Your favorite novel will evolve chapter by chapter, shaped by your reactions. Your game narrative will alter course as your mood shifts. Films will offer you optional endings based on your emotional resilience.

We're not there yet. But we're closer than we think. And as personalization grows more intimate, the very concept of culture begins to fragment.

The Shift: From Shared Culture to Solitary Mythologies

For most of the 20th century, we lived in the age of the monoculture. The Beatles, *I Love Lucy*, Michael Jackson, *Game of Thrones*. These weren't just entertainment—they were rituals. Reference points. Social glue.

Walter Cronkite told you what happened. *Star Wars* gave your generation a mythology. Even divisive icons like Madonna or Kurt Cobain helped define the edges of shared identity.

That era is ending.

AI-generated content—personalized, ephemeral, and infinite—will dissolve the cultural scaffolding that once held us together. Instead of asking, "Did you see the new Marvel movie?" we'll ask, "What did your machine make for you this week?"

In this future:
- Common culture splinters
- Virility becomes niche
- Art becomes a mirror, not a message board

It will feel liberating at first. A renaissance of specificity. A flood of content tailored to your aesthetic soul. But it will come at a cost: we lose the shared symbols that tether us to each other.

What happens to democracy, ritual, and identity when the public square becomes millions of private theaters?

Historical Echoes: From Broadcast to Broadband

This isn't the first time media evolution has reshaped identity.

When Gutenberg's press made books cheap, the world fragmented into literate and illiterate, Catholic and Protestant, and empirical and poetic. When broadcast television arrived, it homogenized culture across nations—suddenly, suburban Ohio and downtown Tokyo had overlapping references. It was the age of the anchor, the mass event, the global premiere.

The internet splintered that again. Social media multiplied the fractures.

But AI is different. This is not just more fragmentation—it is atomization. Where even the *artist* may not exist. The creator becomes the machine, and the audience of one becomes the new norm.

Culture becomes solipsistic. That doesn't mean empty—but it does mean lonely.

So...What Becomes of the Human Artist?

It's tempting to conclude that human artists are obsolete in this new world. But the opposite may be true.

In a sea of synthetic voices, what we crave may be the *real*. Not because it's better—but because it's shared. Because someone else made it—not *for* you, but *despite* you.

In the coming age, human-made art may play four essential roles:

1. Cultural Anchors

Live performance will matter again—not for spectacle, but for solidarity. A campfire. A protest. A spoken-word night in a church basement. These will be our new rituals.

2. Emotional Cartographers

Where AI reflects back what we already feel, human artists will chart the emotional terrain we haven't explored yet. They'll reintroduce us to ambiguity, discomfort, and contradiction—states machines don't do well.

3. Authenticity Signposts

The brushstroke that breaks. The out-of-tune harmony. The lyric that doesn't quite rhyme. These "flaws" will become signatures of the soul. Proof of struggle. Echoes of touch.

4. Machine Whisperers

Many artists won't resist AI—they'll collaborate with it. The best will shape models like clay. They'll say: show me the dream I didn't know I had. Then they'll refine it, ruin it, and remake it—humanize it.

The Museum of Shared Culture (A 2085 Thought Experiment)

Let's imagine a child visiting a museum in 2085.

She walks through a hall titled *The Age of Broadcast Unity*. On display: Beyoncé's 2023 Renaissance tour. The final episode of *Succession*. A looping clip of Will Smith slapping Chris Rock—one of the last collective gasps of the internet age.

"Back then," the docent says, "people watched the same things. Listened to the same songs. It gave them rhythm."

The child tilts her head. "Why would you want that?"

Then she sees a small room. Inside, a single painting. Rough. Asymmetrical. The colors don't match. The brushstrokes are angry. The plaque reads:

Unknown Artist. Not made for anyone in particular.

She stares longer than expected. Then asks, quietly:

"Who was she?"

Philosophical Reckonings: What Is Art For?

This isn't just about algorithms. It's about meaning.

AI can optimize for preference. But preference is not purpose.

A custom playlist doesn't make you weep at your wedding. A procedurally generated love story doesn't stitch up your grief. A machine can give you what you want. Only art can surprise you with what you *need*.

The danger isn't that AI becomes better than artists. It's that it replaces curiosity with comfort.

Rollo May, the existential psychologist, once wrote that "creativity arises from the tension between spontaneity and limits." AI removes the limits. But so does entropy. So does death.

Limits are what give art stakes.

And in that vacuum of friction—where nothing is refused, nothing is resisted—we may forget what made creativity sacred in the first place: that someone *chose* to make it. That they risked embarrassment, failure, and indifference. That they made it not for you, not for anyone, but because they had to.

Conclusion: Meaning, Not Metrics

We will soon live in a world where your AI knows you better than your partner. Where it generates stories that resonate more deeply than your childhood favorites. Where every piece of content has your fingerprint in it—because it was designed to.

And that will be magical.

But what will endure—what must endure—is not optimization, but openness. Not preference, but presence.

The future of creativity is not about replacing the artist. It is about remembering why we ever needed one.

Because at the heart of art—behind the brushstroke, the melody, and the flicker of film—is not a dataset. It's a human. Messy. Mistaken. Miraculous.

And when their story interrupts yours—uninvited, unoptimized, and unrepeatable—that is not noise in the system.

That is meaning.

20 THE ARTIST IN 2035: BETWEEN THE MIRROR AND THE CAMPFIRE

Somewhere, a child is sketching with her finger on a glowing tablet. A melody hums in the background, generated from the emotions she entered into her AI companion's interface. A story begins to unfold, co-written in real time with a prompt engine that knows her taste for open endings and bittersweet minor chords. This is how art begins in 2035.

In this final chapter, we look forward—not to speculate wildly, but to connect the threads we've already seen forming in 2025. Based on current signals, emerging trends, and the historical cycles of creativity and technology, the next ten years are less a leap and more a logical extension. If AI is the new creative partner, what does life look like for an artist a decade from now? And more crucially, why does it still matter?

I. Childhood in the Age of Companion Tools

Artistic inspiration has always had strange beginnings: a favorite song overheard, a worn comic book, a quiet walk in the woods. But the path from spark to skill has changed. In 2035, a child's first brush with creativity is often guided by a smart system—a co-creator that offers feedback, support, and curiosity.

She might not learn to draw from a teacher, but from a voice that asks questions: "Do you want this to feel more like Chagall or Studio Ghibli?" Her first songs are written with assistance from a personal model trained on her family's emotional moments. Her early films are storyboards populated by avatars who respond to her tone of voice.

The world she learns in is more responsive than ever—but also more curated. Taste becomes collaborative from the start.

II. A New Model of Learning

YouTube replaced the classroom for millions of creators over the last decade. In 2035, that learning experience has grown more tailored—and more embedded.

Creative education is now interactive and deeply adaptive. Tools like ProtoSkill and Athena.AI assess a learner's emotional state, prior knowledge, and aesthetic preferences to design curricula on the fly. They suggest references, challenge assumptions, and simulate collaborative feedback. It's not rote instruction; it's dialogic mentorship.

Rather than memorizing history, students train aesthetic models. A young writer doesn't just study Octavia Butler—she uploads the tone of Butler's work into a story assistant that offers genre-bending suggestions. A filmmaker doesn't just emulate Wong Kar-wai—they train a camera model to emulate his pacing and mood.

As a result, "style" becomes both an artifact and a function.

III. How Artists Share Work

In 2025, TikTok and Spotify defined success in terms of reach. But by 2035, mass appeal has given way to micro-audiences. Artists no longer perform for everyone. They perform for the emotionally aligned.

Each user has a personalized media arena—a hybrid of feed, playlist, gallery, and mood board. Artists release works to segmented clusters based on audience memory graphs, emotional profiles, and even real-time biometric feedback. You don't just hear a new artist—you *resonate* with them.

Instead of "going viral," success is measured by depth. Did the work change a mood? Did it generate a conversation? Did it echo?

This doesn't mean public art is gone. But shared experience is rare. Festivals, salons, and IRL events are coveted as communal rituals. They are, in a deeply connected world, moments of real proximity.

IV. Making a Living in the Age of Generative Abundance

The economics of art are unrecognizable.

Artists monetize in layers. One layer is the art itself: songs, poems, and visual works. But increasingly, the second layer is the model—the aesthetic algorithm trained on their choices. Fans don't just buy a painting. They subscribe to a color engine tuned to the artist's emotional scale.

Generative tools allow artists to scale without losing integrity. A single poet might create hundreds of variations of a verse for different emotional states. A songwriter might license their voiceprint for co-creation by fans, then audit and release the best results.

And live performance—unpredictable, unpolished—is gold. Artists who choose to show up physically are celebrated for their risk. One such performer, known simply as Hana, refuses backing tracks and refuses post-production. Her motto: "I want you to know I might fail."

She is, of course, enormously popular.

V. The Fading Shape of Fame

Fame in 2035 is asymmetrical. You might be deeply known to 1,200 people and unknown to the rest of the world. An artist's influence is less about reach and more about fit.

The trade-off? Visibility. Many creators work anonymously, licensing their aesthetic to models that generate far beyond their capacity. They remain unnamed even as their style spreads widely. In some corners, fame is seen as a distraction. The work matters. The self does not.

Still, others chase it. AI tools make fame *look* achievable—synthetic voices can be styled into stardom. Avatars can perform globally. The machinery of virality has never been more accessible.

But meaning? That still costs something.

VI. The Mirror Problem

Here's the risk. With AI models trained on personal data, the art we receive increasingly reflects *us*. Music adapts to our biofeedback. Novels are generated to hit our narrative sweet spots. Visuals match our aesthetic histories.

We may feel more seen than ever. But we may also stop seeing anything new.

This is what philosopher Bernard Stiegler feared: an erosion of shared memory. Each of us becomes the protagonist in a private mythology. And art, once a bridge between perspectives, becomes a mirror that only shows our own face.

This makes the role of the artist more vital than ever.

Not to flatter. To *disrupt*. To say, "You didn't ask for this, but you need to see it."

VII. Ethics and Ownership in the Age of Frictionless Replication

By 2035, synthetic media is indistinguishable from the real thing. A 2025 study already showed that 73% of people couldn't identify AI-generated voices. That number is likely even higher today.

In such a world:

- Can you own a style?
- Can you prove authorship?
- Does the audience even care?

Artists must now watermark their models. Verify their works. Claim their aesthetic fingerprints. But this too is porous. A model can be copied. A style can be cloned.

What becomes valuable, then, is presence.

The artist who shows up. The one who says, "This is me."

VIII. Art as Shared Signal

Despite the noise, art still serves a singular function: to signal what matters. Culture isn't just the content we consume. It's the content we *share*. It's what we remember, what we pass on, what we say, "This moved me—hear it."

That act of choosing together is the glue. Not the algorithm. Not the model. The *act of sharing*.

In a world of infinite personalization, the shared moment becomes revolutionary. A song we both cried to. A film we both debated. A poem that somehow knew us both.

The artists who generate these moments? They matter more than ever. Not because they make the most—but because they *mean* the most.

And somewhere, right now, a child has just told her AI companion: "I want to write a story about a girl who can hear the ocean in people's voices."

The AI offers a suggestion. She reads it. It's pretty good.

Then she puts the tablet down. Walks outside. Listens to strangers talk.

And in that small moment—confused, curious, and unoptimized—an artist is born.

EPILOGUE: THE HAND IN THE LOOP

The machines will keep getting smarter. That much is certain.

Smarter, faster, cheaper—these are the benchmarks by which we've long measured technological progress. Every industrial age is defined not by what it preserves but by what it disrupts. The Jacquard loom. The telegraph. The transistor. The microprocessor. Now, generative AI.

But what makes this era remarkable is not just the relentless ascent of AI. It's the fact that so many of us—trained in slower, stranger, and more tactile methods—are still here. Still showing up. Still reaching for something beautiful, or resonant, or just weird. Not because an algorithm suggested it. But because something in us insisted.

That's the thread running through this transition. Not nostalgia—continuity.

From tape hiss to prompt whisper, from splicing film reels to training neural nets, we don't merely adopt new tools. We reshape them, bending them with the habits, muscle memories, and emotional instincts we've spent a lifetime refining. And in that reshaping, maybe—just maybe—we help the machines glimpse something they'd never find on their own.

Because for all its speed and scope, AI still doesn't know what to want. It can generate endlessly. But it cannot care.

That part's still up to us.

The Typing Pool and the Prompt Whisperer

In the 1940s, my grandmother worked in a typing pool at Fort Myer—she was an army brat and my grandfather was off at war. She was fast, precise, and took pride in typing a page without a single correction. Then came the word processor. It didn't just change a typist's job—it changed the context of writing itself. And yet typing didn't vanish. It adapted, like many of my grandmother's colleagues, who went to taking up roles as editors, administrators, and knowledge workers. They brought with them the grace of precision and the cadence of careful thought.

What we're seeing now with generative AI is a similar leap—but on a more disorienting scale. This time, it's not just the format or the speed that's changed. It's the foundation of authorship.

When a model can write a plausible screenplay in the style of Aaron Sorkin or generate a song that sounds uncannily like Joni Mitchell, we're forced to ask: What is originality, really? And how much of creativity comes from the person behind the curtain—the flawed, stubborn human—versus the pattern itself?

We haven't answered that yet. But many of us keep making anyway. We keep writing, composing, sketching, and tweaking—not in defiance of the machine, but in dialogue with it. Because creativity, at its core, isn't just production. It's provocation. A way of asking questions we don't know how to answer yet.

What Machines Still Miss

In 1959, C.P. Snow (1905–1980)—a British scientist, novelist, and civil servant—warned of a growing divide between "the two cultures," science and the humanities. His concern was that engineers didn't read Shakespeare, and poets didn't grasp thermodynamics. The result, he feared, would be a society blind to its own blind spots.

What he didn't anticipate was a future where artists would wield algorithms and engineers would train models on poetry. A future where the "two cultures" would merge—not seamlessly, but productively—into a messy, generative third space.

And yet, there's still a gap. Because while AI can simulate fluency, it still lacks agency. It can complete a sentence or mimic a style, but it cannot decide—independently—to take a risk, to tell the truth, to lie, to say something because it simply *has to be said.*

When John Coltrane blew past the chord changes or Joan Didion left an idea hanging at the end of a paragraph, they weren't optimizing for coherence. They were chasing the unknown. They were betting on the emotional force of a choice no machine could rationalize. That's not a style. That's a decision. And machines, for all their pattern-matching power, still don't make decisions. They output probabilities. We imbue them with meaning.

When New Tools Arrive

The arrival of photography in the 1830s sparked panic among painters. Why render the world by hand, they asked, when a camera could do it with greater fidelity and far less time?

But the panic gave way to reinvention. Painters embraced abstraction, impressionism, symbolism—styles no camera could replicate. Technology didn't extinguish their art. It freed it.

The same thing happened with recorded music. When phonographs and later synthesizers entered the scene, critics predicted the death of live performance and acoustic instruments. Yet from that friction emerged jazz, sampling, hip-hop, and ambient. New genres were born from tools the old world feared.

Today's moment with AI feels similar—but more accelerated. Because the machine is no longer just an instrument. It's a collaborator. A co-writer. A remix engine.

The question isn't whether AI can generate music or writing or code. It can. The question is whether the output carries any weight beyond its technical achievement.

Some of the most compelling uses I've seen treat AI not as a shortcut, but as a strange mirror. A choreographer in Tokyo runs pose-estimation models on dancers, then choreographs around the glitches. A poet in Nairobi trains a small language model on Swahili proverbs, then feeds it surreal prompts. In each case, the human uses the machine not to automate expression, but to provoke it. That's not outsourcing. That's art.

The Ghosts in the Dataset

Behind every well-trained model is a dataset filled with human voices—some canonical, some anonymous. The machine doesn't know the cost of those voices. It doesn't know the loneliness in Sylvia Plath's syntax or the blood behind Nina Simone's phrasing. It only knows the shape of the signal.

That's where ethical friction lives. When we train machines on the work of the dead without their consent—or on the work of the living without compensation—we risk flattening cultural history into a resource to be mined, not a lineage to be honored.

This is not an argument against AI. It's a reminder: when we speak of training data, we are speaking of people. Of minds. Of scars and revelations encoded in text, brushstroke, and breath.

To lose sight of that is to misread what we've built. These machines don't create. They remember us—badly, sometimes beautifully. What they reflect is what we feed them. What they distort is what we forgot to value.

The Value of Doing Things the Hard Way

There's something deeply subversive about making things by hand in the age of automation. A handwritten letter. A song written on an acoustic guitar. A sketchbook filled with bad ideas and breakthroughs.

These aren't acts of resistance. They're reminders that friction is part of meaning. That speed is not the same as depth. That convenience can be the enemy of care.

In the coming years, there will be pressure—from platforms, from markets, and from algorithms—to produce more. To optimize, adapt, and comply. But some of the most important work will come from those who pause. Who complicate. Who ask why before they ask how.

AI will keep evolving. It will pass more tests, write more convincing scripts, generate more plausible images. But it still won't know what it's doing. It will still lack the one thing that art—real art—requires.

A reason.

The Last Word

The machines will keep getting smarter. But the more urgent question isn't what they'll become.

It's what we'll become, alongside them.

We are not just teaching machines to think. We are also re-teaching ourselves what matters. That includes the value of human ambiguity, the unpredictability of risk, and the weird, stubborn need to make something for no reason other than the fact that it *feels* right.

Because for all its power, AI still doesn't know what to want.

It doesn't grieve. It doesn't wonder. It doesn't ache. It doesn't *care*.

That part—the spark, the motive, the mystery—that's still up to us.

About the Author

Michael-Patrick Moroney is an award-winning technologist, music producer, and creative strategist who's spent decades at the intersection of art and technology. He's helped launch some of the first digital music platforms (ArtistOne, AWAL), built platforms, campaigns, branding and content for brands like iTunes, Microsoft, Google, IBM, BMW, Amnesty International and Virgin Records, and collaborated with artists from indie legends to global superstars. His writing and consulting explore how emerging technologies—especially AI—are reshaping creativity, identity, and culture. He lives in upstate New York with his wife, designer Cara Cragan, a very slim British Labrador, four Main Coon cats, and more guitars than he needs.

www.ingramcontent.com/pod-product-compliance
Lightning Source LLC
Chambersburg PA
CBHW071155200326
41519CB00018B/5241